Advances in Biometrics

G.R. Sinha
Editor

Advances in Biometrics

Modern Methods and Implementation
Strategies

 Springer

Editor
G.R. Sinha
Electronics and Communication
Engineering
Myanmar Institute of Information
Technology
Mandalay, Myanmar

ISBN 978-3-030-30438-6   ISBN 978-3-030-30436-2   (eBook)
https://doi.org/10.1007/978-3-030-30436-2

This Springer imprint is published by the registered company Springer Nature Switzerland AG.
The registered company address is: Gewerbestrasse 11, 6330 Cham, Switzerland

Dedicated to my late grandparents, my teachers, and Revered Swami Vivekananda

Preface

Biometrics is made up of two words, "bio" and "metrics," which mean the techniques using biological features or human characteristics for authentication or identification of a person. Biometrics is the most commonly searched and used technique, and the field of biometrics has been attracting a number of novel advancements, right from driverless cars to all automated and AI-based systems. IoT and wearable devices are hugely using biometrics for authentication or identification purpose, and thus, this emerging area of signal processing has become the most important component in all research and advancements in the world. Biometrics focuses on using physical and behavioral traits such as fingerprint, speech, face, iris, DNA, palm print, and many others. There are several intricacies in the field of biometrics that need to be properly addressed with potential applications and case studies. Some of these issues include uniqueness of features, template protection, robustness, social implications, cognitive computing, 3D face detection, privacy, future directions, etc.

This book covers almost all major aspects of modern applications of biometrics: Introduction to Biometrics and Special Emphasis on Myanmar Sign Language Recognition; Handling the Hypervisor Hijacking Attacks on Virtual Cloud Environment; Proposed Effective Feature Extraction and Selection for Malicious Software Family Classification; Feature-Based Blood Vessel Structure Rapid Matching and Support Vector Machine-Based Sclera Recognition with Effective Sclera Segmentation; Different Parameter Analyses of Class1 Generation2 (C1G2) RFID System Using GNU Radio; Design of Classifiers; Social Impact of Biometric Technology Myth and Implications of Biometrics, Issues, and Challenges; Segmentation and Classification of Retina Images Using Wavelet Transform and Distance Measures; Language-Based Classification of Document Images Using Hybrid Texture Features; Research Trends and Systematic Review of Plant Phenotyping; Case Studies on Biometric Application for Quality-of-Experience Evaluation in Communication Services; Nearest Neighbor Classification Approach for Bilingual Speaker and Gender Recognition; and Dimensionality Reduction and Feature Matching in Functional MRI Imaging Data.

The purpose of this book is to bring out fundamentals, background, and theoretical concepts of biometrics in comprehensive manner along with their potential applications and implementation strategies. There are various methods of biometrics reported and highlighted invariably by few authors and researchers. This book covers case studies, real-time applications, and research directions in addition to basic fundamentals. The book will be very useful for wide spectrum of target readers such as research scholars, academia, and industry professionals, especially for those who are working biometric methods, related issues, challenges, and problems.

Mandalay, Myanmar   G.R. Sinha

Acknowledgment

I express my sincere thanks to my wife, Shubhra; my daughter, Samprati; and my great parents for their wonderful support and encouragement throughout the completion of this important book *Advances in Biometrics: Modern Methods, Implementation and Strategies*. This book is an outcome of focused and sincere efforts that could be given to the book only due to great support of my family.

I am grateful to my teachers who have left no stones unturned in empowering and enlightening me, especially Shri Bhagwati Prasad Verma who is like godfather for me. I extend my heartfelt thanks to Ramakrishna Mission order and Revered Swami Satyaroopananda of Ramakrishna Mission Raipur India.

Sincere thanks to all the contributors for writing relevant theoretical background and real-time applications of biometric methods and entrusting upon me as editor.

I also wish to thank all my friends, well-wishers, and all those who keep me motivated in doing more and more, better and better.

My reverence with folded hands to Swami Vivekananda who has been my source of inspiration for all my work and achievements.

Last but most important, I express my humble thanks to Michael McCabe, Senior Editor, *Applied Sciences* of Springer Nature for his great support, necessary help, appreciation, and quick responses. I also wish to thank Arun Pandian for his support and entire team of Springer Nature for giving me this opportunity to contribute on some relevant topic with reputed publisher.

Contents

Editor's Biography

G.R. Sinha is Adjunct Professor at the International Institute of Information Technology (IIIT), Bangalore, and currently deputed as Professor at Myanmar Institute of Information Technology (MIIT), Mandalay, Myanmar. He obtained his B.E. (Electronics Engineering) and M.Tech. (Computer Technology) with Gold Medal from the National Institute of Technology, Raipur, India, and his Ph.D. in Electronics and Telecommunication Engineering from Chhattisgarh Swami Vivekanand Technical University (CSVTU), Bhilai, India.

He has published 223 research papers in various international and national journals and conferences. He is active Reviewer and Editorial Member of more than 12 reputed international journals such as *IEEE Transactions on Image Processing*, Elsevier's *Computer Methods and Programs in Biomedicine*, etc. He has been Dean of the Faculty and Executive Council Member of CSVTU India and currently a Member of Senate of the MIIT. He has been appointed as ACM Distinguished Speaker in the field of DSP for years (2017–2020) and as Expert Member for Vocational Training Program by Tata Institute of Social Sciences (TISS) for 2 years (2017–2019). He has been Chhattisgarh Representative of IEEE MP Sub-Section Executive Council for the last 3 years. He has served as Distinguished Speaker in Digital Image Processing by Computer Society of India (2015) and as Distinguished IEEE Lecturer in IEEE India council for Bombay section. He has been Senior Member of IEEE for the last many years.

He is Recipient of many awards like TCS Award 2014 for Outstanding Contributions in Campus Commune of TCS; R. B. Patil ISTE National Award 2013 for Promising Teacher by ISTE New Delhi; Emerging Chhattisgarh Award 2013; Engineer of the Year Award 2011; Young Engineer Award 2008; Young Scientist Award 2005; IEI Expert Engineer Award 2007; ISCA Young Scientist Award 2006 Nomination; and Deshbandhu Merit Scholarship for 5 years. He has authored six books including *Biometrics* published by Wiley India, a subsidiary of John Wiley and Medical Image Processing published by Prentice Hall of India. He is Consultants of various skill development initiatives of NSDC, Government of India. He is Regular Referee of project grants under DST-EMR scheme and several other schemes of the Government of India. He has delivered many keynote/invited talks and chaired many technical sessions in international conferences in Singapore, Myanmar, Bangalore, Mumbai, Trivandrum, Hyderabad, Mysore, Allahabad, Nagercoil, Nagpur, Kolaghat, Yangon, Meikhtila, and many other places. His special session on "Deep Learning in Biometrics" was included in IEEE International Conference on Image Processing 2017. He is Fellow of IETE New Delhi and Member of international professional societies such as IEEE, ACM, and many other national professional bodies like ISTE, CSI, ISCA, and IEI. He is Member of various committees of the university and has been Vice President of Computer Society of India for Bhilai Chapter for 2 consecutive years. He has guided 7 (08) PhD scholars and 15 M.Tech. scholars. His research interest includes image processing and computer vision, optimization methods, employability skills, outcome-based education (OBE), etc.

Chapter 1
Introduction to Biometrics and Special Emphasis on Myanmar Sign Language Recognition

G.R. Sinha and Pyae Sone Oo

1.1 Introduction and Background

Biometrics combines two important words, bio and metrics, and thus biometric deals with some biological measures or metrics employing biological features. Biometric concepts, theories, fundamentals, and applications become very relevant in present context since almost all technological advancements are using some type of biometric techniques for various reasons, such as authentication, security, etc. For example, in cell phone fingerprint biometrics is commonly used to authenticate right user/owner of phone to avoid any misuse of the functionalities of the phone. For example, when a person buys a new phone, that time the user is asked to register his/her fingerprint impressions which are recorded in the form of some suitable representation, and when authenticated user puts the fingertip while using (testing process in biometrics language), then the person is authenticated as the right person. Sinha et al. in their book on biometrics highlight the concepts and several emerging applications of the biometrics [1]. A typical block diagram of any suitable biometric method which depicts important flow of processes in biometrics can be seen in Fig. 1.1.

Actually, Fig. 1.1 shows a general purpose block diagram of biometrics; in fact it involves two important stages, namely, training and testing. Training has the following components:

- Image capturing or acquisition: Images such as faces, iris images, and finger-prints are captured in this process.

G.R. Sinha (✉) · Pyae Sone Oo
Electronics and Communication Engineering, Myanmar Institute of Information Technology, Mandalay, Myanmar
e-mail: gr_sinha@miit.edu.mm; 2015-MIIT-ECE-006@miit.edu.mm

© Springer Nature Switzerland AG 2019
G.R. Sinha (ed.), *Advances in Biometrics*,
https://doi.org/10.1007/978-3-030-30436-2_1

Fig. 1.1 Biometric process

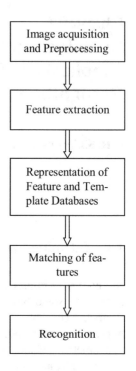

- Preprocessing: This involves several steps for making the images suitable for further stages, such as resizing, reformatting, de-noising, etc. De-noising or image enhancement is also an important task that deals with any unwanted signal added in original images.
- Feature extraction: In this process, suitable set of features are extracted from the images that were captured and preprocessed.
- Representation of features: This process converts features of images into some suitable representations, generally referred to as templates, and the database of templates is created.

Testing stage involves all processes of training and few more, which are:

- Template matching: The face or any image which is being tested for recognition or authentication is subjected to all processes of training, and finally the template extracted from the test face or image is matched against all those which are saved in template database. If the image is matched properly, then it is authenticated.
- Recognition: After matching, the testing stage results in either authenticated or not. The classification or matching tasks are done by a number of algorithms and methods available. In fact, now it is very difficult to choose appropriate method for matching process since robustness is a big issue in the field of biometrics.

Actually, the recognition is also of two types, authentication and verification. In verification, the test face is verified as the authorized user, whereas as in authentication, the user is recognized exactly who the user is, and therefore verification is also known as 1 to N matching, and authentication is referred to as one to one matching. For example, in banking application, if an unauthorized user attempts to enter in the secured premises or attempts to make use of secured applications, then by using face or fingerprint biometrics, the intruder can be detected and caught.

Historical study of biometrics suggests that the concept of biometrics is very old and was used in the beginning of the nineteenth century for criminal investigation. There were traditional methods of using biometrics, either using some features of palm, fingers, or so, but now the technology has made the use and application of biometrics so sophisticated that the technology is available on microscale and nanoscale in numerous applications such as robotics, computer vision, DNA matching, protein synthesis, medical image analysis, remote sensing, and satellite imaging, and the list is huge. Sinha et al. study the role of biometrics in understanding the ability of human brain employing deep learning as a training method [2]; and one more such study is reported in [3] by Sinha et al. Sinha et al. implemented a signal language recognition using their own sign language databases as Indian sign language (ISL) database and implemented Devanagari text and numeral recognition using different methods, and this work proves to be extremely useful for speech- and vision-disabled people [4]. Patil et al. (2016) and Patil et al. (2011) implemented ISL and Devanagari character recognition using shift invariant feature transform (SIFT) and hidden Markov model (HMM) [5, 6]. Snehlata et al. in their work implemented multimodal biometrics using principal component analysis (PCA) method and achieved satisfactory performance in comparison with unimodal biometrics [7, 8]. Here, multimodal biometrics involved multiple modalities or traits such as face, iris, fingerprint, palm print, etc., whereas in unimodal the recognition is only on the basis of single modality, and the recognition accuracy is always limited due to challenges and problems with the modality chosen.

Handwritten recognition is researched extensively in [9–12] using SIFT- and HMM-based methods. One interesting application of biometrics is in license plate recognition (LPR) of vehicle number plates which is useful in automatic parking, surveillance, and toll systems. The research on LPR is found on number plates of a number of countries and implemented differently because the number plates and standards are different from country to country. Choubey et al. (2013) and Choubey et al. (2011) implemented bilateral portioning method and pixel count method for LPR that dealt very well with confusing number plates in case of presence of noise, such as I and T and O and Q [13, 14]. Lazrus et al. employed neural network for Indian number plate recognition [15], and the research studies were carried out for Myanmar number plate recognition (New Ni et al. 2018) in [16]. Indian license plate recognition was also implemented using some hybrid approaches, like neuro-fuzzy method in [17, 18] by Siddhartha et al.

1.2 Classification

The biometric methods are classified on the basis of several parameters such as types of traits, number of traits, and the authentication type. The methods are divided on the basis of types of modalities, as follows:

- Face recognition: Faces are biometric modalities or traits.
- Iris recognition: Iris of eyes is taken as trait.
- Fingerprint matching: Fingerprint impressions are modalities used in biometrics.
- Palm print matching: Palm prints are traits here.
- Hand geometry recognition: Based on using hand movements.
- Signature recognition: Considered as behavioral biometrics because signature is considered as one of the important behavioral traits in human being.
- Speech recognition: Uses speech as modality.
- Facial expression recognition: Faces with different mood conditions of a person such as sadness, happiness, laughter, etc. are taken into consideration for determination the type of nature or behavior of person.

Classifications on the basis of number of modalities are:

- Unimodal biometrics: Involves single biometric trait such as only face in simple face recognition.
- Multimodal biometrics: Involves more than one modality, such as AADHAAR car, which is very popular and widely used authentication card issued to every citizen in India which includes the face, iris, and all ten fingers of a person.

The way authentication is performed, biometrics is also classified:

- Authentication: one to one matching
- Verification: one to many matching

A different way of classifying the biometrics results:

- Soft biometrics: This utilizes some unique and non-transferable features of persons such as permanent scar mark, mole, etc.
- Hard biometrics: All other biometrics except soft are called as hard biometrics.

Initial work on biometrics suggests that most of biometrics are based on fingerprint matching. One such work is reported in [19] that highlights basic overview and types of biometrics giving stress on fingerprint recognition. This paper presents advantages of fingerprint and its working principle also, exploiting features of fingerprints and minutiae which are some representations of fingerprint ridges and valleys. The market-wise potential is also largest in fingerprint biometrics since it is used most commonly and popularly in enormous number of applications [1]. Fernando et al. in the most recent work discuss about selfie biometrics as a kind of face biometrics. Super-resolution methods are implemented and compared with iris recognition of large number of samples [20]. Nowadays, biometrics is being used in travel and other identity documents; in all travel modes especially in air travel, attempt is made to free the passengers from carrying boarding passes, and they can be allowed only on the basis of biometric system.

1.3   Societal and Ethical Issues

Implementation of biometrics at large scale for benefit of masses requires user participation at all levels, right from biometric capturing process to user-end support. When the implementation is targeted as large masses, then it has some limitations related to social implications and ethical matters. For example, AADHAAR is a most common identity tool in India, and every citizen in India has been issued this card [1], but the implementation of the project in terms of data capturing of faces, iris, and fingerprints raised few important issues in society, and many of experts were of opinion that the biometrics is violation of personal secrecy and privacy. The parliaments have discussed a number of times on this matter, and a number of amendments are made to address the issues raised as concerns from various groups or the society.

In a technical report on biometrics [21], social implications were discussed in detailed with few case studies. The European Commission (2005) in this report highlighted various types of biometric technologies with implementation strategies, diffusion of methods, and focused on DNA, face, and iris biometrics. Among many social aspects [1], few important are reported here:

- Social exclusion and human factors: Ethnicity, age, gender, etc. are also needed in declaration process of biometric applications used for common mass beneficiary schemes. More amount of research on user-friendliness and usability of biometric data need to be carried out so that awareness can be created and people can support the implementations, since there are some inhibitions or apprehensions in providing the data, while training process is performed in biometric application.
- Feeling of trust breach: Sometimes, the user participation is also affected due to the lack of trust between users (citizens) and implementation authority.
- Privacy of personal information: Various private and public sectors take the personal data of citizen for deploying and providing biometric services that causes apprehension in common masses with a feeling of their personal data being stolen and might be misused also.

Jain et al. also presented overview of social acceptance and challenges in biometrics [22]. In financial transactions such as credit card usage, health insurance, and other similar areas of market and commercial potentials, apprehensions are obvious of misuse or breach of privacy related to biometric personal data.

1.3.1 Ethical Issues

In technical report [21], ethical issues related to biometrics were also elaborated with examples. Biometric data can be used by various law enforcement agencies for various legal procedures that again require personal data, though law enforcement systems are devising laws for protection of personal data, but still the matter becomes serious when it involves personal data in legal system and delivery [1]. Emilio et al. studied social and ethical implications in biometrics particularly and was suggested that the implications and issues may be different for different nationalities, but some issues are unavoidable [23]. A number of reports mentioned how the reports attempted to address few ethical issues, such as RAND report (2001) addressed sociocultural issues; the European Commission report (2005) attempted to address social and ethical issues. In the report, physical privacy, religious objections, and personal information were said to be at risk or misused by the service providers or application facilitators. A wide range of legal, medical, and social issues were introduced in several reports [23].

1.4 Soft Biometrics

Traditional biometrics involves common modalities such as faces, fingerprints, iris images, hand geometry, facial expression, etc., but there are few biometric or biological traits which are unique and non-transferable, for example, scar, tattoos, weight, mark on face or somewhere else, mole, etc. [1]. Reid et al. use soft biometrics in application for surveillance where tattoos, body geometry, and scars were taken into consideration for biometric identification [24]. In a video footage from CCTV capture, the soft biometric traits are detected and made the basis of identification using some metrics for matching like false accept (FA), false reject (FR), and equal error rate (EER); and the main basis of matching between test samples against the database is Euclidean distance. Other suggested semantic traits as soft biometric modalities include arm length, chest, hair color, arm shape, leg shape, leg length, etc. One of the advantages of soft biometrics was reported as continuous authentication is not affected with time change since other modalities may change with time but some of soft traits do not change over time [1, 24].

Antitza et al. presented a survey on soft biometrics and suggested that the soft traits which are referred to as ancillary traits such as gender, age, scar, hair, color, and weight are used in combination with primary modalities such as face and fingerprints in order to improve reliability and matching accuracy of biometric system. This can be easily done in multimodal system by fusing both types of biometric traits [25]. Main advantages of the soft biometrics are:

- Human can understand the attributes related to soft traits easily.
- Robustness is reported to be better in implementing the methods for matching.

- These modalities can be taken or captured without consent for certain application, and thus privacy issues are not much important.
- Easier way of getting taxonomy of modalities, such as demographic, medical, behavioral, and geometric.

A number of methods which are used for soft biometric recognition were summarized in terms of their major findings and database size [25]. Few of them include support vector machine (SVM), neural network, and principal component analysis (PCA). There are some datasets also reported in this survey like CASIA gait DB [25] and IRIP gait DB [25].

1.5 Biometric Standards, Protocols, and Databases

Deployment of biometrics and services are required to follow certain standards or policies and also need to work under a set of rules and procedures, referred to as standards and protocols [1]. Generally, the standards are developed for:

- Supporting biometric applications and interoperability
- Conforming various architectures where biometrics are deployed and used
- Suggesting common metrics and models for use
- Dealing with usability, quality, and interoperability for biometrics applications
- Testing and research and development facilitation

Frances discussed the standards, norms, and protocols [26] in details with regard to a popular biometric application in India where each citizen has been given a unique ID (identity) also referred as UID (unique identity) that includes three modalities [1]. This policy papers recommend several suggestions and policies for implementing similar biometric for other nations as UID which is also called as AADHAAR card in India. Actually, the implementations like UID can save huge amount of money where a large number of ghost workers are claimed in various projects; and it has been benefitting India to great extent, and now no one can weed out money in the name of ghost workers since each account is being associated with UID number and the money directly goes to their account as direct benefit transfer (DBT) to the beneficiaries. This report highlights difficulty in such issues in Cambodia, Tanzania, and Nigeria where a lot of money is weeded out just for the reason that the system like UID is not working. If the UID like implementation has to spread across many places, it has to follow certain rules and standards. The department that takes care of UID project, UIDAI, has initiated certification process for biometric equipment to be used for applications especially those dealing for a large number of masses. The certification is done through standardization testing and quality certification (STQC) which has set of procedures for testing, evaluation monitoring of various biometric equipment, and deployment.

UIDAI committee has framed certain biometric standards [27], biometric design standards for UID applications (2009), which include standards for faces, fingerprints with best practices, and members of different sub-committees. There are some important documents which were used in designing the standards:

- ISO/IEC 1544 for JPEG 2000 image coding
- IAFIS-IC-0110 (V3) for fingerprint image compression
- ISO/IEC 19785-1:2006, ISO/IEC 19794-2:2005, ISO/IEC 19794-4:2005., ISO/IEC 19794-5:2005, ISO/IEC 19794-6:2005, and ISO/IEC CD 19794-6.3, respectively, for data specification, minutiae data, fingerprint data, face data, and iris data as Part 1, Part 2, Part 4, Part 5, Part 6.

1.5.1 Standards

The main objective of designing standards is to make interoperability easier and compatible between biometric devices and supporting IT systems. There are few important agencies involved in design process, such as American National Standard Institute (ANSI), European Committee for Standardization (CEN), International Organization for Standard (ISO), International Committee for Information Technology Standards (INCITS), and open and advancing standards for information society (OASIS).

Standards for face images include photographic requirements, enrollment, source type, pose, image compression, format, and feature blocks. ISO/IEC 19794-5 covers all such requirements for capturing face images for biometric applications. To sum up about these set of procedures and norms [1, 27]:

- Image being captured should of very good quality, probably best quality.
- Full frontal images of 300 dpi and 24 bit RGB color space should be captured.
- Expression of face needs to be without smile, mouth closed, and open eyes.
- Roll, yaw, and pitch angles are required to be within $5°$ on positive as well as negative value.
- JPEG 2000 image compression to be used.

Similarly, fingerprint image standards ISO/IEC 19794-4 [27] covers the following:

- 500 dpi images for enrollment with 8-bit pixel depth and 200 gray levels of dynamic range.
- 500/300 dpi for authentication.
- JPEG2000 compression scheme should be used.

All other standards discussed have their own procedures and norms, for minutiae extraction and storing, image compression, etc. The standards for iris images in UIDAI, ISO/IEC 19794-6 include recommendations of some sub-committees, for example, ISO/IEC JTC 1/SC 37, introduced in 2010 [27]. Technical details, biometric accuracy, and best practices for each type of modalities are explained in detail [27]. International Telecommunication Union (ITU) suggests several organizations which are involved in designing and developing standards for various purposes related to biometric applications [28]. ISO and CEN are of those international organizations mainly responsible for development of such standards; one more such organization is International Electrotechnical Commission (IEC). NIST is an example of national-level organization for development of standards; few more are there, for example, Bureau of Indian Standards (BIS) and American National Standard Institute (ANSI). Industrial consortia are also there like IEEE that has separate consortium for this purpose as IEEE biometric consortium. There are numerous committees and their sub-committees for technical standards developments [28], such as:

• ISO/IEC JTC 1/SC 37 for biometric standards
• ISO/IEC JTC 1/SC 17 for standards related to personal identification
• ISO/IEC JTC 1/SC 27 for standards of security applications in IT systems

Electronic identity in Peru is DNIe (digital national identity) and Estonia has ID-Kaart as identity card [28], and all of them follow some standards governed by International Organization for Standardization, their committees, and sub-committees. Biometric standards and databases are discussed in [29] for visiting the USA.

1.5.2 Protocols

Protocols are used for an important task of dividing the databases into a number of datasets. For example, we have a huge face database that involve faces of different poses, different illuminations, and different face and head positions, and then the database can be portioned into number of datasets based on poses, head positions, illumination, etc. This process of partitioning is referred to as protocol which is actually some set of rules responsible for partition [1, 30–32]. Anongporn presents his doctoral thesis on authentication protocols used for biometric applications [33] and suggested ProVerif model and CPV02 model. Comparison of various types of models used as protocols was discussed in terms of different types of databases portioned into a number of datasets. Focus was made on security protocols used for authentication purpose.

1.5.3 Databases

Initially, in UID work, three databases were created as DB1, DB2, and DB3 [27]. DB1 covers 27 urban and 81 rural areas including 1351 images and single impression sensor technology. This employed FIPS 2001 APL standard and also image quality specifications of FBI. DB2 includes 20,000 persons and 200K total images all fingerprints segmented properly and prepared for database. DB3 has 5600 individuals that contain 56,000 images approximately [27]. This example of databases for biometrics is what was used in UID project, and all biometrics applications have their own databases separately for faces, fingerprints, and iris images like NIST has its own databases satisfying various biometric standards. Few other databases are described in [30] by Poh et al. in their report on biometric testing and evaluation, and such repositories are:

- US-based biometric consortium.
- CASIA is a database of center for biometrics and security research China.
- NIST databases.
- AR face databases of Spain.
- AT&T databases of faces.
- BANCA, a multimodal database.
- CASE-PEARL face database.
- FERET databases of NIST.
- JAFFE, Japanese female face database of facial expressions.
- KFDB, Korean face database.
- Yale Face database.
- MCYT-DB, fingerprint database.
- FVC 2000, FVC 2002, FVC 2004, and FVC 2006 all fingerprint databases.
- ICE databases for iris images by Iris Challenge Evaluation.
- CASIA-Iris-V3 for iris images.
- Many others as multimodal biometric databases for vein and hand geometry.

Melissa [34] and Aglika et al. also focused on databases study and survey [35]. Coding scheme was suggested in [35] for designing and implementing the protocols in multimodal biometric databases. Asma et al. studied extensively on multimodal biometric databases [36].

1.6 Myanmar Sign Language Recognition

As discussed in the beginning, biometrics is most commonly used in a number of security- and authentication-based applications. Based on modalities, the applications are classified. Speech recognition and face recognition are common biometrics among all biometric methods. Speech biometrics falls under behavioral biometrics, and face recognition is considered as physiological type of biometrics [1]. One more important application of biometrics is sign language recognition which uses hand geometry, palm geometry or movement which is very useful for people with hearing and vision disability. Sign language, as its name suggests, helps the disabled persons by communicating with the help of some gestures, symbols, and expressions. These movements and gestures may be captured from moving images (videos) as well as static images [1]. This section focuses on Myanmar sign language biometrics with implementation; sample results and discussion juts to showcase a case study of biometrics.

1.6.1 Sign Language Recognition

Sign language biometrics or recognition is researched in many countries as per their requirement, culture, and varying symbolic gestures [1, 37–55]. Thad et al. studied American Sign Language using video captured from wearable device [37]. Hidden Markova model (HMM) was used and tested in MIT research laboratory. Sana also implemented American Sign Language (ASL) using Otsu segmentation method [38]. Helen et al. discussed about sign language recognition using linguistic sub-units. The system used three types of sub-units for consideration which are learnt from appearance data and 2D and 3D tracking data. Then the sub-units are combined into a sign-level classifier with two options. HMM and sequential pattern boosting are used that provided 54% and 76% accuracy, respectively. The work was tested for big datasets having 984 signs, and it was observed that as the number of signs increases, then the accuracy is adversely affected [39]. Assaleh et al. presented sign language for Arabic language using HMM and resulted 94% of accuracy [40]. Razieh et al. implemented a multimodal biometrics for sign language recognition using Restricted Boltzmann Machine (RBM). The system achieved about 90% accuracy, but there is some difficulty on recognizing characters with low visual interclass variability such as high hand post similarity [41]. Irene implemented Iris sign language recognition [42]; Patil et al. on Indian Sign Language (ISL) in [43], Brandon et al. on ASL using neural network [44], Wijayanti et al. on alphabet sign language [45], Yanhua et al. [46] on Japanese Sign language Biometrics [46], Joyeeta et al. on ISL using video signal as input [47], Deepali et al. on ISL biometrics [48], Amit et al. on ASL [49], Paulo et al. on Portuguese Sign Language recognition [50], Cao et al. on ASL [51], Alina et al. on sign language biometrics [52], Dominique et al. on sign language recognition [53], and Fitri et al. on alphabet sign language recognition are few major research contributions in the field of sign language recognition [54]. Though there are several research works on sign language biometrics, most of them suffer from lack of robust method and robust databases.

1.6.2 Myanmar Sign Language (MSL) Recognition and MIIT Database

Though there are numerous research on sign language biometrics, the research in this area is limited. In fact, the number of research on MSL is almost insignificant. However, there are significant amount of work in natural language processing on

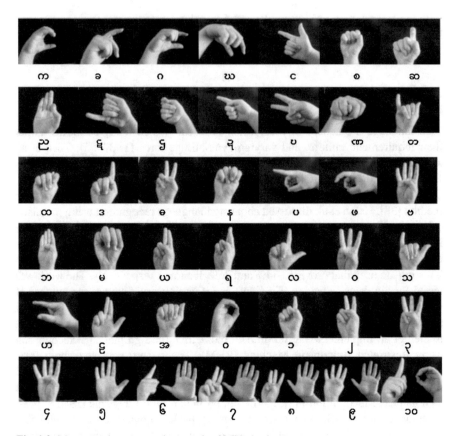

Fig. 1.2 Myanmar characters and numerals of MSL database

Myanmar language, grammar, and related analysis. One such work on MSL is reported in [55] by Thit et al. on Myanmar sign language text analysis using genetic algorithm which presents an overview of MSL biometrics.

We have attempted successfully to implement MSL using standard method and shift invariant feature transform (SIFT) that operates on databases of MSL created by our research group. Figure 1.2 shows the database developed by our research group working at MIIT Mandalay Myanmar toward their undergraduate project course.

The databases include arrays of alphabets and numbers ["Ka Gyi" "Ka Kway" "Ga Nge" "Ga Gyi" "Nga" "Sa Lone" "Sa Lane" "0" "0" "Nya" "Ta Talin Jade" "Hta Won Bell" "Dain Yin Gouk" "Dain Yin Hmote" "Na Gyi" "Ta Won Bu" "Hta Sin Htoo" "Da Dway" "Da Out Chike" "Na Nge" "Pa Sout" "Pha Oo Htote" "Ba Htet Chike" "Ba Gone" "Ma" "Ya Pa Lat" "Ya Gout" "La" "Wa" "Tha" "Ha" "La Gyi" "Ah" "0" "1" "2" "3" "4" "5" "6" "7" "8" "9" "10"]. The symbols are recorded as gestures of palm by the MIIT research team and recorded audio file also for each number and alphabet. The symbolic representations of the characters were verified with the help of a school in Mandalay running for deaf and dumb people. Deaf charity Mandalay Myanmar helps in collecting database and other gestures were recorded. All 33 alphabets (from "Ka" to "Ah") and 11 numbers ("0" to "10") are created and recorded as images. We also created sound or audio files for all these data with the help of Google Translate Text to Speak (TTS) assistant. Then, some preprocessing was applied to resize and reformat the images.

1.6.3   MSL Implementation and Results

The SIFT algorithm [1] is implemented for recognition of Myanmar Sign Language. SIFT extracts key points for matching, and while testing, key points are matched against those stored in database of key points. Key points may be considered here as template representation [1]. The steps of SIFT can be simply interpreted by the following:

- An input image of any gesture can be recognized with comparison of the image against databases of the images. The images are not stored as images but kept inside template database.
- Euclidean distance is used to match the images based on nearest value using feature vector stored inside the database. Feature vector is only the template here.
- SIFT has its meaning due to inclusion of scale, location, and orientation value which are extracted from the images and thus making the method appropriate and can take any orientation of input image.
- Clusters of feature vector are determined using Hough transform.
- Features are of two types, detector and descriptor. Frames are extracted by detector having some variations. Descriptor connects the regions and makes association to the images so that features become invariant of shift, scaling, rotation, orientation, and illumination variation.
- The images that were subjected to the method are gray scale images in order to reduce the computation time and save memory. Thus, all input images captured originally as color images are converted into gray scale before they are subjected to various stages of MSL recognition.

Figure 1.3 shows the steps in concise manner where preprocessing, initialization of databases, and creation of feature vector are important components of MSL biometrics.

The detailed flow diagram for implementation of MSL biometrics is shown in Fig. 1.4 where almost all stages and their functionalities are self-explanatory. If more than one result is produced, then threshold value or Euclidean distance is again checked and evaluated until we get a single and decisive output.

Now, results for an image are shown here to highlight the working of the method for MSL biometrics and Myanmar database. Figure 1.5 shows an original image from the MIIT database, and Fig. 1.6 is the grayscale image of Fig. 1.5.

SIFT frames and peak points can be seen in Figs. 1.7 and 1.8.

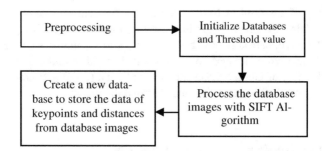

Fig. 1.3 Main stages in MSL recognition

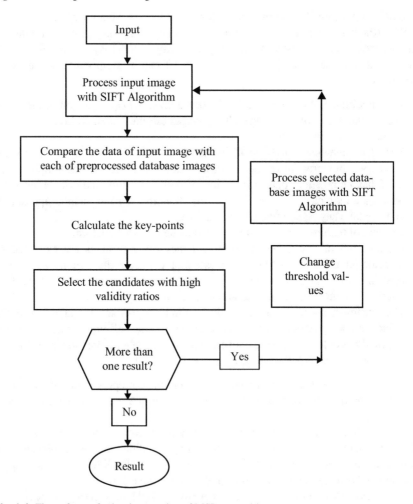

Fig. 1.4 Flow of steps for implementation of MSL recognition

Fig. 1.5 Original image as input

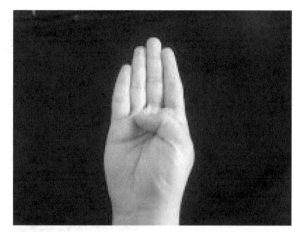

Fig. 1.6 Gray scale image of input

Fig. 1.7 SIFT frame

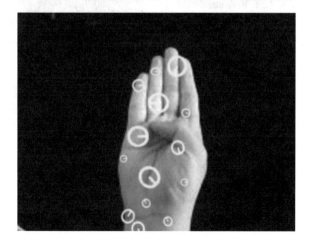

Fig. 1.8 Test SIFT peak
threshold parameter

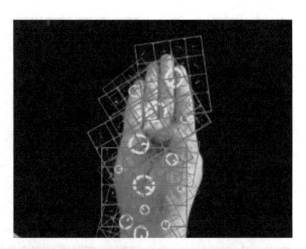

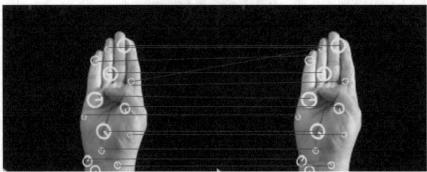

Fig. 1.9 Matching between features in two images

Fig. 1.10 Peak threshold 0

Fig. 1.11 Peak threshold 10

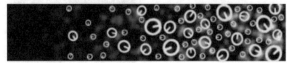

Descriptors help in finding similar regions in two images on the basis of matching key points in test image against image in database, as can be seen in Fig. 1.9.

Detector controls two values, peak threshold and edge threshold, as shown in Figs. 1.10, 1.11, 1.12, and 1.13. The number of frames that were detected for different threshold values are shown in figures.

Similarly, the results highlighting the number of frames for different values of edge threshold are shown in Figs. 1.14, 1.15, 1.16, and 1.17.

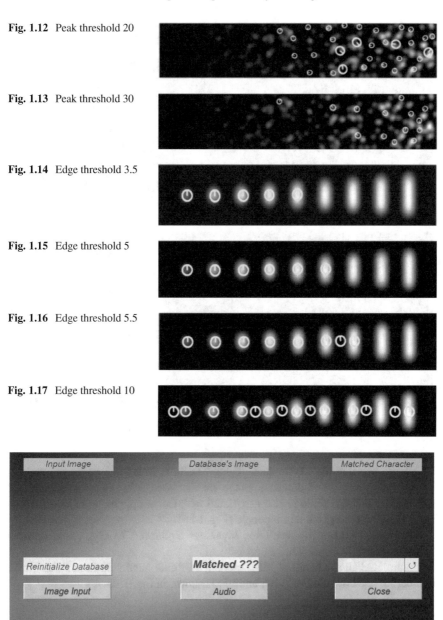

Fig. 1.12 Peak threshold 20

Fig. 1.13 Peak threshold 30

Fig. 1.14 Edge threshold 3.5

Fig. 1.15 Edge threshold 5

Fig. 1.16 Edge threshold 5.5

Fig. 1.17 Edge threshold 10

Fig. 1.18 A GUI for testing

Now implementation of MSL biometrics on MATLAB platform is briefly presented with the help of GUIs obtained while implementing and testing the work for Myanmar Sign Language recognition. Figure 1.18 shows a GUI for testing the work, and an input image is given in the system as can be seen in Fig. 1.19.

Fig. 1.19 An image given as input

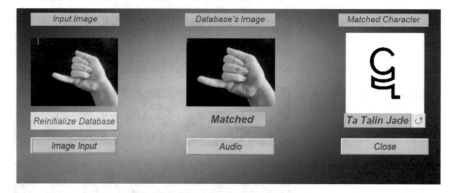

Fig. 1.20 Myanmar alphabet "Ta Talln Jade" recognized with suitable audio

We can see that the Myanmar alphabet "Ta Talln Jade" is recognized, and the images being tested and database image are the same. Most important feature of this work is that the audio is also associated with the sample being matched which can greatly help deaf and dumb people (Fig. 1.20).

Figure 1.21 shows the result for Myanmar alphabet "Ka Gyi," and alphabet "Nga" is shown in Fig. 1.22.

Figure 1.23 shows result of MSL biometrics for alphabet "Sa Lane." Result for number "1" is shown in Fig. 1.24, and "7" is shown in Fig. 1.25.

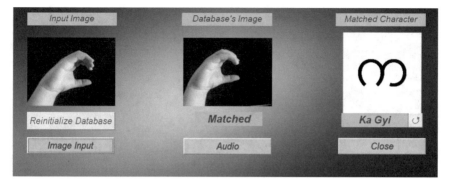

Fig. 1.21 Myanmar alphabet "Ka Gyi" recognized with suitable audio

Fig. 1.22 Myanmar alphabet "Nga" recognized with suitable audio

Fig. 1.23 Myanmar alphabet "Sa Lane" recognized with suitable audio

Fig. 1.24 Myanmar number for "1" recognized with suitable audio

Fig. 1.25 Myanmar number for "7" recognized with suitable audio

1.7   Conclusions

The present chapter discussed overview, history, and background information of biometric techniques. Biometric databases, protocol, and standards were also presented with a number of real-time examples. An emphasis is given to Myanmar Sign Language recognition which has worked well with validation of results in terms of sound associated with each symbol or gesture.

Acknowledgments The authors acknowledge the support of Deaf Charity Mandalay Myanmar and Ma Su Aeindray Htet, a student of department of Computer Science and Engineering of MIIT Mandalay for her support and help while discussing with Charity in Mandalay.

References

1. S. Patil, G.R. Sinha, *Biometrics: Concepts and Applications* (Wiley India Publications, a subsidiary of John Wiley, New Delhi, India, 2013)

2. G.R. Sinha, K.S. Raju, R. Patra, D.W. Aye, D.T. Khin, Research studies on human cognitive ability. Int. J. Intell. Defense Support Syst. **5**(4), 298–304 (2018)
3. G.R. Sinha, Study of assessment of cognitive ability of human brain using deep learning. Int. J. Inf. Technol. **1**(1), 1–6 (2017)
4. G.R. Sinha, Indian sign language (ISL) biometrics for hearing and speech impaired persons: Review and recommendation. Int. J. Inf. Technol. **9**(4), 425–430 (2017)
5. S.B. Patil, G.R. Sinha, Distinctive Feature Extraction for Indian Sign Language (ISL) Gesture using Scale Invariant Feature Transform (SIFT). J. Inst. Eng. India Ser. B **98**, 19–26 (2017). https://doi.org/10.1007/s40031-016-0250-8
6. B.P. Sandeep, G.R. Sinha, V.S. Patil, Isolated Handwritten Devnagri numeral recognition using HMM: A Holistic Approach. Second International Conference on Emerging Applications of Information Technology, pp. 185–189, 19–20 February, 2011, Kolkata, India
7. S. Barde, A.S. Zadgaonkar, G.R. Sinha, PCA based multimodal biometrics using ear and face modalities. Int. J. Inform. Technol. Comput. Sci. **5**, 43–49 (2014)
8. S. Barde, A.S. Zadgaonkar, G.R. Sinha, Multimodal biometrics using face, ear and iris modalities. Int. J. Comput. Appl. **2**, 9–15 (2014)
9. S.B. Patil, G.R. Sinha, Real time handwritten Marathi numerals recognition using neural network. Int. J. Inf. Technol. Comput. Sci. **2**(8), 76–81 (2012)
10. S.B. Patil, G.R. Sinha, K. Thakur, Isolated handwritten Devnagri character recognition using Fourier descriptor and HMM. Int. J. Pure Appl. Sci. Technol. **8**(1), 69–74 (2012)
11. S.B. Patil, G.R. Sinha, Feature extraction and matching using scale invariant Feature transform for Indian sign language. International Conference on Science and Technology for Sustainable development, pp. 5012–5019, 26 May 2016, Nashik
12. S.B. Patil, G.R. Sinha, Intensity based distinctive feature extraction and matching using scale invariant feature transform for Indian sign language, Mathematics and Computers in Science and Engineering Series. Proceedings of the 17th International Conference on Mathematical Methods, Computational Techniques and Intelligent Systems Mathematics and Computers in Science and Engineering Series, Mathematical Methods and Systems in Science and Engineering, Tenerife, Spain, ISBN: 978-1-61804-281-1, p. 245–255, 10–12 Jan 2015
13. S. Choubey, G.R. Sinha, A. Choubey, Bilateral Partitioning based character recognition for Vehicle License plate. International Conference on Advances in Information Technology and Mobile Communication – AIM 2011 April 21–22, 2011, Nagpur, Maharashtra, India, V.V. Das, G. Thomas, and F. Lumban Gaol (Eds.): AIM 2011, CCIS 147, pp. 422–426, 2011, © Springer-Verlag Berlin Heidelberg 2011
14. S. Choubey, G.R. Sinha, License Plate Localization using novel recursive algorithm and Pixel Count Method. i-manager's J. Embed. Syst. **2**(1), 06–16 (2013)
15. A. Lazrus, S. Choubey, G.R. Sinha, An efficient method of vehicle number plate detection and recognition. Int. J. Mach. Intell. **3**(3), 134–137 (2011)
16. N.N. Kyaw, G.R. Sinha, K.L. Mon, License plate recognition of Myanmar vehicle number plates A critical review. 2018 IEEE 7th Global Conference on Consumer Electronics (GCCE 2018), pp. 771–774, Japan
17. S. Choubey, G.R. Sinha, Pixel distribution density based character recognition for vehicle license plate. 2011 International Conference on Network and Computer Science (ICNCS 2011), 5, pp. 26–30, Kanyakumari India, 2011
18. S. Choubey, G.R. Sinha, B.C. Patel, A. Choubey, K. Thakur, Pixel clustering based partitioning technique for character recognition in vehicle license plate. 2011 3rd International Conference on Machine Learning and Computing (ICMLC 2011), 4, pp. 390–394, 26–28 Feb 2011, Singapore
19. H. Richard, An introduction to biometrics and large scale civilian identification. Int. Rev. Law. Comput. Tech. **13**(3), 337–363 (1999)
20. F. Alonso-Fernandez, R.A. Farrugia, J. Fierrez, J. Bigun, Super-resolution for selfie biometrics: Introduction and application to face and iris, in *Selfie Biometrics: Methods and Challenges*, (Springer, 2019), pp. 1–24. http://gigantes.ii.uam.es/fierrez/files/2019_HBookSelfie_SuperSelfieFaceIris_Alonso.pdf

21. Technical Report Series on Biometrics at the Frontiers: Assessing the Impact on Society, For the European Parliament Committee on Citizens' Freedoms and Rights, Justice and Home Affairs (LIBE) European Commission Joint Research Centre (DG JRC), Institute for Prospective Technological Studies, 2005, https://www.statewatch.org/news/2005/mar/Report-IPTS-Biometrics-for-LIBE.pdf
22. A.K. Jain, A. Ross, S. Prabhakar, An introduction to biometric recognition. IEEE Trans. Circuits Syst. Video Technol. **14**(1), 4–20 (2004)
23. E. Mordini, C. Petrini, Ethical and social implications of biometric identification technology. Ann. Ist. Super. Sanita. **43**(1), 5–11 (2007)
24. D. Reid, S. Samangooei, C. Chen, M. Nixon, A. Ross, Soft biometrics for surveillance: an overview, in *Machine Learning*, (Theory and Applications. Elsevier, 2013), pp. 327–352. https://eprints.soton.ac.uk/342219/
25. A. Dantcheva, P. Elia, A. Ross, What else does your biometric data reveal? A survey on soft biometrics. IEEE Trans. Inf. Forensics Secur. **11**, 441–467 (2015)
26. F. Zelazny, The evolution of India's UID program: lessons learned and implications for other developing countries, CGD Policy Paper 008. Center for Global Development, Washington, D.C., 2012, http://www.cgdev.org/content/publications/detail/1426371
27. Biometrics Design Standards For UID Applications, Version 1.0, UIDAI Committee on Biometrics Unique Identification Authority of India, 2009, http://www.corporatelawreporter.com/wp-content/uploads/2013/05/Biometrics-Design-Standards-For-UID-Applications.pdf
28. Technical Standards for Digital Identity, Draft for discussion among World Bank Group, http://pubdocs.worldbank.org/en/579151515518705630/ID4D-Technical-Standards-for-Digital-Identity.pdf
29. Biometric Standards Requirements for US-VISIT, United States Visitor and Immigrant Status Indicator Technology (US-VISIT) Program, Version 1.0 March 15, 2010, https://www.dhs.gov/xlibrary/assets/usvisit/usvisit_biometric_standards.pdf
30. N. Poh, C.-H. Chan, J. Kittler, J. Gabbaly, J. Fierrez, A Anjos, S. Marcel, C. Karabat, Biometrics evaluation and testing, Funded under the 7th FP (Seventh Framework Programme, Theme SEC-2011.5.1-1, https://www.beat-eu.org/papers/d83.pdf
31. C. Braz, A security usability protocol for user authentication, Doctoral thesis project, University of Quebec at Montreal. Concordia University, 2007
32. Draft Biometric Open Protocol Standard, P2410/D11, February 2015, IEEE-SA Standards Board, Copyright © 2014 by The Institute of Electrical and Electronics Engineers, Inc. 20 Three Park Avenue, New York, USA, https://www.oasis-open.org/committees/download.php/56664/P2410d11.pdf
33. A. Salaiwarakul, Verification of secure biometric authentication protocols, Ph.D. thesis, School of Computer Science, The University of Birmingham, United Kingdom, June 2010
34. M. Chase, O. Dunkelman, Privacy preserving biometric database, http://www.cs.haifa.ac.il/~orrd/crypt/biometric.pdf
35. A. Gyaourova, A. Ross, A coding scheme for indexing multimodal biometric databases, Proceedings of IEEE Computer Society Workshop on Biometrics at the Computer Vision and Pattern Recognition (CVPR) conference, (Miami Beach, USA, 2009), https://www.cse.msu.edu/~rossarun/pubs/RossMultimodalIndexing_CVPRW2009.pdf
36. A. Kebbeb, M. Mostefai, F. Benmerzoug, C. Youssef, Efficient multimodal biometric database construction and protection schemes. int. Arab. J. Inf. Techn. **12**(4), 346–351 (2015)
37. T. Starner, J. Weaver, A. Pentland, Real-time American sign language recognition using desk and wearable computer based video. IEEE Trans. Pattern Anal. Mach. Intell. **20**(12), 1371–1375 (1998)
38. S.K. Jadwa, Otsu segmentation method for American sign language recognition. Int. J. Eng. Res. Gen. Sci. **3**(5), 916–923 (2015)
39. H. Cooper, N. Pugeault, R. Bowden, Sign language recognition using sub-units. J. Mach. Learn. Res. **13**, 2205–2213 (2012)
40. K. Assaleh, T. Shanableh, M. Fanaswala, F. Amin, H. Bajaj, Continuous Arabic sign language recognition in user dependent mode. J. Intell. Learn. Syst. Appl. **2**, 19–27 (2010)

41. R. Rastgoo, K. Kiani, S. Escalera, Multi-modal deep hand sign language recognition in still images using restricted Boltzmann machine. Entropy **20**(809), 1–15 (2018)
42. I. Hernandez, Automatic Irish sign language recognition, Thesis of Master of Science in Computer Science (Augmented and Virtual Reality), University of Dublin, Trinity College, 2018
43. S. Patil, G.R. Sinha, Offline mixed Devanagari numerals recognition using artificial neural network. Adv Comput Res **4**(1), 38–41 (2012)
44. B. Garcia, B. Garcia, Real-time American sign language recognition with convolutional neural networks, pp. 225–232, 2016, http://cs231n.stanford.edu/reports/2016/pdfs/214_Report.pdf
45. W.N. Khotimah, R.A. Saputra, N. Suciati, R.R. Hariadi, Alphabet sign language recognition using leap motion technology and rule based back propagation-genetic algorithm neural network. Jurnal Ilmiah Teknologi Informasi **15**(1), 23–37 (2017)
46. Y. Sun, N. Kuwahara, K. Morimoto, Analysis of recognition system of Japanese sign language Using 3D Image Sensor, 2013, http://design-cu.jp/iasdr2013/papers/1159-1b.pdf
47. J. Sinha, K. Das, Recognition of Indian sign language in live video. Int. J. Comput. Appl. **70**(19), 17–22 (2013)
48. D. Kaushik, A. Bhardwaj, Hand gesture recognition on Indian sign language using neural network. Int. J. Innov. Eng. Technol. **6**(4), 554–565 (2016)
49. A.K. Gautam, A. Kaushik, American sign language recognition system using image processing method. Int. J. Comput. Sci. Eng. **9**(7), 466–471 (2017)
50. P. Trigueiros, F. Ribeiro, L.P. Reis, Vision-based Portuguese sign language recognition system, 2014, https://www.researchgate.net/publication/262698351_Vision-Based_Portuguese_Sign_Language_Recognition_System
51. C. Dong, M.C. Leu, Z. Yin, American sign language alphabet recognition using Microsoft Kinect, 2015, https://ieeexplore.ieee.org/abstract/document/7301347
52. A. Kuznetsova, L. Leal-Taix´e, B. Rosenhahn, Real-time sign language recognition using a consumer depth camera, 2013, https://ieeexplore.ieee.org/document/6755883
53. D. Uebersax, J. Gall, M. Van den Bergh, L. Van Gool, Real-time sign language letter and word recognition from depth data, 2012, https://ieeexplore.ieee.org/document/6130267
54. F. Utaminingrum, K. Somawirata, G.D. Naviri, Alphabet sign language recognition using K-nearest neighbor optimization. J. Comput. **14**, 63–70 (2019)
55. T.T. Zaw, Myanmar text classifier using genetic algorithm, M. C. Sc. thesis, University of Computer Studies Yangon, Myanmar, 2018

Chapter 2
Handling the Hypervisor Hijacking Attacks on Virtual Cloud Environment

Su Su Win and Mie Mie Su Thwin

2.1 Introduction

Early on year 2000, virus writing was so popular in all ICT ranges. The enormous amounts of virus took place in the field of cyber domain, and millions of hosts are infected. The name of the virus, so-called the Sober virus, infected and spread out over 218 million machines in 7 days. The email virus, Mydoom, was sent to about 100 million of machine as an infected email. The I LOVE YOU virus affected 55 million of machines by gathering usernames and passwords. These attacks were just about boasting, not to earn money.

In the last decade, virus writing is old-fashioned or out-of-date. Then, it was the time of malware and phishing attacks. They were hacker's intentional target for financial purpose by deceiving every individual to get username and password from innocent users.

After that, new-generation, cloud environment will easy to use for the people. Conventional management and control platforms are countering huge challenges concerning with security. So, next-generation systems will require to be more dependent and flexible for more secure cloud environment. The cloud computing architecture model can solve the problem associated with resource utilization, allocation, managements, etc.

However, many elements for example, well-styled architecture of cloud system are still not flexible and difficult to modify, adapt, and change to be associated with this fashion. It consists of crucial network topologies, many components, and

S. S. Win (✉)
Information and Communication Technology Research Centre (ICTRC), Yangon, Myanmar

M. M. S. Thwin
Cyber Security Research Lab, University of Computer Studies, Yangon, Myanmar

© Springer Nature Switzerland AG 2019
G.R. Sinha (ed.), *Advances in Biometrics*,
https://doi.org/10.1007/978-3-030-30436-2_2

dimensions of the user control over infrastructure as a service, platform as a service, or software as a service.

Nowadays, otherwise, in current modern era, we have to move our private information and data from local workstations and servers to cloud computing architecture where all of these data are very attractive and live behind the cloud service provider. So the game of the hacking process is changing and encounters new technical challenges.

Behind the scenes of the cloud computing is virtualization infrastructure. The meaning of virtualization is a construction of virtual (rather than actual) machines which can run multiple operation systems on a single PC or device, but it has to share all hardware resources, for instance, servers, storage devices, network devices, operating system, desktop, and applications virtualization.

The main idea of virtualization is to share IT properties and resources in order to get benefits from abstraction of layers in business, organizations, and all government sectors. The physical machine which creates the virtual machine is referred to as host machine and is also known as a guest machine.

One standard example in virtualization environment is hypervisor. A hypervisor is located in a place between the virtual machines (VMs) and the real physical hardware device. By using this kind of software, layer abstraction separation provides a great chance for system admin to be more flexible when managing and controlling virtual machines.

However, because of the common connection between VMs and physical layer, hypervisor can be regarded as a risk carrier when compromising and propagating of threat and risk. Unlike physical network architectures, it is difficult to see log, countermeasure function like penetration testing (network pen test) and scanning.

The complexity of virtualization with a new challenge is hyper-jacking; the new-born hyper-jacking revolves around the business world's emerging enthusiasm for application, operating system, and issue of virtualization. Hyper-jacking expresses the hypervisor stack jacking. Hyper-jacking involves setting up a rogue hypervisor that can acquire complete control of a server.

Hyper-jacking or hypervisor attack is a great approach not only compromising a server and stealing data but also in maintaining the persistence. As soon as it is getting control of the hypervisor, it can control everything running on the machine. The hypervisor is the single point of failure (SPOF) in cyber security, and if it is lost, protection of sensitive information may also be lost. This increases the degree of risk and exposure may be large. Today, the conventional security monitoring and measure tools are inadequate to harden the operating system, so machine can be compromised. Figure 2.1 shows before installation hypervisor on PC and after attack on a hypervisor.

Although cloud computing has plenty of benefits with virtualization, it brings numerous amount of security vulnerabilities. This paper presents three objectives: firstly, to understand the terms of virtualization vulnerabilities in cloud computing; secondly, to recognize the virtualization threats; and, thirdly, to give the mitigation technique and awareness knowledge for hypervisor attacks and hyper-jacking-style threats that include a particular type of malware called virtual machine-based

Fig. 2.1 Illustration of
layer-wise hypervisor attack

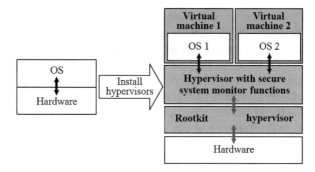

rootkits and also revel avoiding methods with behavior-based hypervisor detecting
method.

2.2 Related Works

Till now, central processing unit (CPU) of Intel and AMD processor is vulnerable,
and it is difficult to find out with build-in discovery tool to make countermeasure
secure hypervisor handling [1]. So this is an important research topic of security
issue, and all researchers have to try to solve this kind of challenges. The proposed
system reveled exploration, classification, and analysis of vulnerabilities and types
of attacks in virtualization environment by using open-source detection software.

The first and foremost of rootkit that happen on hypervisor is started in 2006
and developed by Joanna Rutkowska. The name of the rootkit is so-called Blue Pill
[1]; it is the first, real, and effective hypervisor rootkit that used driver based on
windows utilized in AMD central processing unit [2]. Similarly, at the same time,
in 2006, Dia Zovi was developed MAC OS and Intel central processing unit, and
Vitriol. Then, detection on hypervisor started in 2007 [1].

Fannon was presented and made analysis in 2014 to prove that the comparison
between two hypervisors [3]. Vitriol and Blue Pill became prominent tools in
information technology (IT) security environment and had persuaded the formation
of many different hypervisor detection methods and approaches.

The exposure of detection technique can be categorized into four groups:
behavior-based, detection-based, signature-based on the trusted hypervisor, and
time-based detecting analysis. The technique with signature-based detection utilized
memory scanning of hypervisors' design and pattern. The other three categories are
based on the interaction and collaboration with a hypervisor [3].

Blue Pill [4] and Vitriol [5] were famous development projects in recent years,
when system is in a run time stage that installs and puts in malicious code to the
hypervisor. At the primary stage, hypervisor exists as an international standard
machine; the above mentioned two projects are able to insert a malicious code to
hypervisor on a memory where there is no need to reboot process.

The system of hacking explains the fast improvement in the new programs that make the codes, offering a better security to the system with more efficiency. The expression cracker also belongs to the same field; it makes use of the hacking skills for the unlawful purposes like email id, intruding into other system. Hacking has different types such as backdoor hacking, viruses and worms, Trojan horses, denial-of-service attack, anarchists, crackers, kiddies and ethical hacking [1].

Hyper-jacking attack controls on a hypervisor to compromise the instance VMs. As a consequence of this kind of attack, the hacker takes over the management and control of the guest operating within VM environment; virtualization server and the host OS will be stilled active and in use. Traditional security mitigation techniques are not adequate because the security measures on the guest VMs or sever do not know that host operating system itself has been compromised. If hyper-jacking is achieved by an attacker, he needs to have a processor that can do hardware-assisted virtualization to access the host. Attacker may convince the admin or user to install some malicious code to attack the hypervisor.

2.3   Background Theory of Proposed System

2.3.1   Virtualization Concept

In the era of information technology, the fundamental change is happening in cloud computing with virtualization; this means that the combination (mixture) of hardware and software engineering process that creates virtualization and it can run on the same platform with multiple operating systems (Fig. 2.2).

Fig. 2.2 Illustration of layer abstraction in virtualization

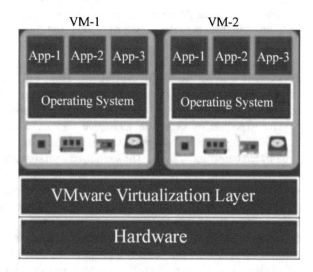

Virtualization process becomes critical for business and organization to seek better resource providing, less hardware, easy IT management with economically. But virtualization is a complicated scheme and continually evolves with certain risks concerning hypervisor security. The concept of virtual machine allows the following functions such as isolation, server consolidation, portability, application portability, suspend, and restart.

Based on the data center technology, types of virtualization can be classified below:

- Hardware virtualization
- Network virtualization
- Storage virtualization
- Server virtualization
- Operating system virtualization
- Desktop virtualization
- Application virtualization [6]

2.3.2 Detection Method Based on Behavior Approach

Detection method based on behavior technique only depends on the system activity, and it can be categorized into two parts: the system with hypervisor and system that starved off a hypervisor. There are three kinds of detection method based on behavior technique shown in (Fig. 2.3). Translation look aside buffer TLB-based detection and methods based on errors in hypervisors and errors in CPUs [7].

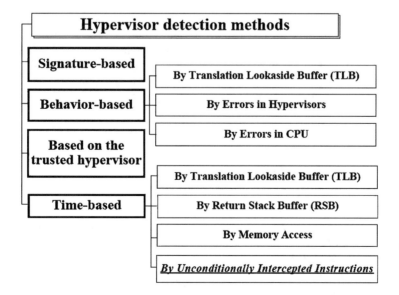

Fig. 2.3 Classification of hypervisor detection methods

2.3.2.1   Translation Lookaside Buffer TLB-Based Approach

Translation lookaside buffer detection approach is likely to be applied at caching memory that stored and used to get more address translation speed to sense a hypervisor [1].

Translation lookaside buffer detection approach includes a set of virtual and physical address; it may be recently accessible to the system. Whenever the operating system has to get access to the memory, translation lookaside buffer entry is looking for this corresponding address. If needed, requested virtual address is exist in the Translation Lookaside Buffer, they saved and retrieved physical address to contact memory.

VM exit guide shows the translation lookaside buffer when a hypervisor is present. On the other hand, without hypervisor, such permission and authorization cannot occur. This is because detection hypervisor reduced checking translation lookaside buffer approach content; this can be done by numerous alternate behaviors, for example, by editing or modifying page table entry (Myers and Youndt 2007) [3].

But, translation lookaside buffer detection approach does not work on central processing units of AMD and other Intel central processing units. The new extra additional translation lookaside buffer fields ASID and process-context identifier will not allow VM exit (virtual machine exit) flush level translation lookaside buffer (TLB) [3].

2.3.2.2   CPU-Based Detection Approach

By using the help of bugs and some instructions in convinced central processing unit model, a hypervisor can be detected whether it (rootkit) is present or not. Other kind of bugs like VMSAVE 0x67 that also freezes the system too. The prefix run time and execution of the VMSAVE 0x67 can halt the virtualization system. These error and bugs occur with hypervisor. If without hypervisor, this kind of error cannot occur (Barbosa 2007). This detection method can be found and applicable in obsolete central processing unit and requires nontrivial adaptation to new central processing units [3].

2.3.3   Hypervisor Model

Hypervisor is a software that distributes and shares computer resources (for instance, processing power unit, random access memory, storage, etc.) in virtual machines, which can be communicated to other computers in the network. It creates virtualization layer that makes server virtualization possible and offers people to share resources. So, the users have to run applications without heavily relying on powerful desktop computers that are costly. Moreover, system administrators can

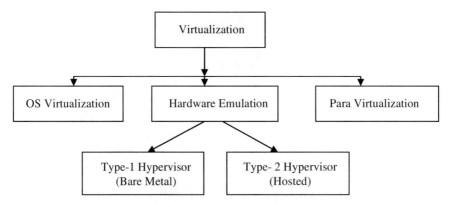

Fig. 2.4 Types of virtualization design chart

also use the hypervisor to monitor and manage VMs with Virtual Machine Manager (VMM). There are two types of hypervisors, and the following are some examples of hypervisors [1] (Fig. 2.4):

- VMware ESX/ESXi
- Hyper-V
- VMware Workstation
- Oracle Virtual-box
- Fusion
- Virtual Server
- Xen Server

2.3.3.1 OS Virtualization

According to Vangie Beal [8] assumption, operating system virtualization refers to the use of software by allowing system hardware to run multiple instances; operating system lets the users to execute different applications.

2.3.3.2 Hardware Emulation

Hardware emulation is generally used to debug, fix, check, and verify a system under enterprise design plan. An administrator has to use hardware emulation if he needs to run an unsupported operating system (OS) within a virtual machine (VM). In such a scenario, the virtual machine does not have direct access to server hardware [9]. Nowadays, virtualization technology can available both free or commercial use. For instance, VMware ESXi, VMware VMs server, Microsoft virtual server, and Xen Server.

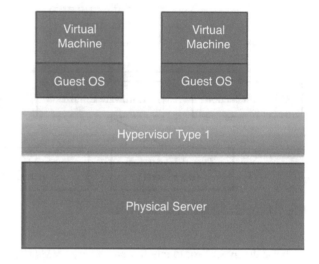

Fig. 2.5 Example demonstration of type 1 hypervisor model

Type 1 Hypervisor

Type 1 hypervisor is loaded directly to hardware; Fig. 2.5 shows the type 1 hypervisor and the following are the kinds of type 1 hypervisors (Fig. 2.6):

- VMware ESX/ESXi for VMware vSphere
- Hyper-V for Microsoft
- Xen Server

Type 2 Hypervisor

On the other hand, type 2 hypervisor is loaded in an operating system running on the hardware that is our laptop or desktop, for example (Fig. 2.7):

- VMware workstation
- Oracle Virtual-box
- Virtual Server
- Fusion for Mac OS

2.3.3.3 Para-Virtualization

Para Virtualization is more complex than hardware emulation technique that mentioned above. It multiplexes access and administrates to hardware infrastructure resources, offering great performance and requiring guest OS modification before deployment, for example, Xen (open source) [8].

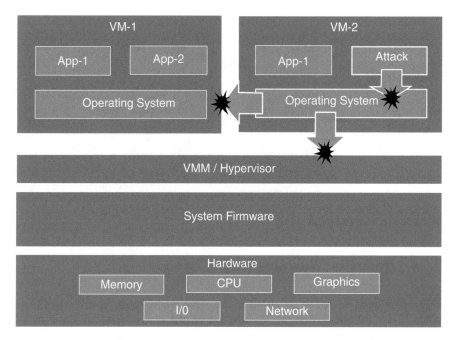

Fig. 2.6 Stack of bare metal hypervisor attack

Fig. 2.7 Example
demonstration of type 2
hypervisor model

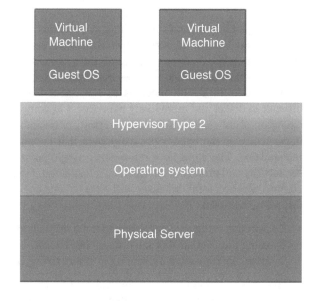

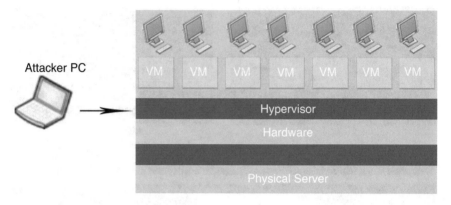

Fig. 2.8 Hyper-jacking attack model

2.3.4 Hypervisor Hijacking Thread Types

Hyper-jacking is a kind of attacking in the place where a hacker takes malicious control over the hypervisor that creates the virtual environment within a virtual machine (VM) host. The idea of that attack is to target the operating system that is below that of the virtual machines so that the attacker's program can run and the applications on the VMs above it will be completely oblivious to its presence [10] (Fig. 2.8).

Hyper-jacking includes installing a malicious activity, fake hypervisor can manage and control to accomplish the entire server system. Regular security measurements are ineffective to secure the system because the operating system is not aware that the machine has been compromised. In hyper-jacking, the hypervisor mainly works in stealth mode is to run under the virtual machine; it is more difficult to detect and more likely gain access to computer servers which affect the operation of the entire institution, company, or business organization. If the hacker gains access to the hypervisor, everything that is connected to that server can be manipulated. The hypervisor represents a single point of failure when it comes to the security and protection of sensitive information. For a hyper-jacking attack to be completely successful, a hacker will have to take control of the hypervisor by the following methods:

- By injecting a rogue hypervisor under the original hypervisor
- By using direct control of the primary hypervisor
- By attacking rogue hypervisor on top of an existing hypervisor definition concept

Hypervisor vulnerability is defined that if hackers manage and achieve to compromise hypervisor software, they will release access to every VM and the data stored on them. While hypervisors are overall well-protected and robust (strong), security specialists can say that hackers will finally discover a bug in the software like a zero-day attack.

Currently, Reports of hypervisor hijacking attacks are very rare and limited; but in concept assumption, cybercriminals can run a program that breaks out of a VM and has direct interaction to the hypervisor. From this step, they can control everything, from access privileges to computing resources. There are three types of hypervisor threats:

- Internal threats
- Technology threats
- External threats

The VMs and hypervisor allocate as a distributed nature. So, another point of weakness in vulnerability is the network. Since hypervisors distribute VMs via the business organization network, they can be susceptible to remove intrusions and denial-of-service attacks if we don't have the right protections in that position. These are some types of common attack vectors [11].

- Virtual CPUs
- Software memory management units
- Interrupting and timer mechanisms
- Input and output (I/O) and networking
- Virtual network layer (e.g., vSwitch)
- System calls or hypercalls
- Hypervisor add-ons or extension

These are some recently happened threats types:

- Vulnerability from outdated operating system and lack of active patching
- Poor network performance (every VM can have a resource cost)
- Intel disclosed Spectre-like L1TF vulnerabilities, August 17, 2018
- Hardware debug documentation leads to widespread vulnerability, May 11, 2018
- AMD patches in testing with ecosystem partners, May 04, 2018 (Fig. 2.9)

In any fields of the system, there are pros and cons, opportunities and challenges. In virtualization technology, it has many aspects of IT management, but the nature of virtualization may also have complicated task of cyber security control in a new threat vector point of view.

2.4 Serious Vulnerabilities in Virtualization

The following sections are phenomena of risk of virtualization that occur when the number of virtual machines on a network reaches a place where the administrator cannot control and manage these VMs effectively. The session shows some attacks with virtual machine and their serious, critical vulnerabilities and so on.

VM architecture and setting are variations of common threats such as denial-of-service attacks, session hijacking attack, DNS hijacking attack, others still hugely

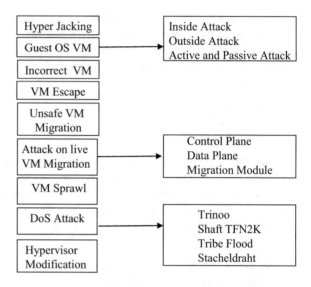

Fig. 2.9 Taxonomy of virtual machine vulnerability to persuade threat

exist as a theoretical but are likely approaching as buzz and means increase. That kind of critical weaknesses are discussed as follows.

2.4.1 VM Sprawl

Virtualization technology is not only improves hardware efficiency but also reduces time and cost. VM sprawl is one of the biggest issues that facing many business organizations using desktop and server virtualization. VM sprawl can occur many VMs on a network where administrator lost control on his virtual machine. Attackers can get opportunities because of lack of systematic monitoring resources [12].

2.4.2 Hyper-jacking Attack

Hypervisors are hijacked to gain access and control of VMs and its data. Hyper-jacking attacks can occur both type 1 and type 2 hypervisor. Although type 2 hypervisors run over a host OS, type 2 hypervisors can be regarded as a theoretical approach because this type of hyper-jacking is very rarely found in real-world virtualization environment due to difficulty of direct access to hypervisor. However, this type of attack can be considered as a real threat, and administrator should plan and prepare for it as an offensive nature [13].

2.4.3   VM Escape Method

In order to get direct access to the hypervisor, guest operating system escapes from their VM encapsulation. This kind of opportunity gives the attacker for all access to VMs, if guest and host privileges are high. This kind of attack is also not well-known by all attackers, but administrators and experts have to consider VM escape as the most serious threats in VM security [13].

2.4.4   Denial-of-Service Attack

An attempt to deprive victim's resource to make unavailable for authorized users by shutting down the services. The attackers targeted different resources such as network resources (data store, CPU and servers) [13].

2.4.5   Incorrect VM Isolation

VM isolation has an important role to make virtualized environment safe. For secure and right sharing resources, VMs have to be isolated from each other. When communication between one VM to another VM, they should be restricted just like a traditional physical machine with physical firewall. On the other hand, because of the poor control and security policies, it can lead to the isolation breaches in VMs. Attackers can exploit and cause incorrect VM isolation which can reduce the VM performance [13].

2.4.6   Unsecured VM Migration or VMotion

This kind of event occurs when VM migrated to a new host machine with lack of security policies and configuration updated. Then, host and other guest operating systems are more vulnerable, and attackers can get more chance to attack because of no awareness of system's weaknesses, vulnerabilities, and alerts by system administrators [13].

2.4.7   Host and Guest Vulnerabilities

Because of multiple weaknesses of windows and operating system, some host and guest communications have several system vulnerabilities in virtualization environment. Similar to that, other systems can lead to vulnerabilities such as in email, web browsing, and network protocol. But co-hosting and virtual linkages can make serious attack and cause VM damaging effect [13].

2.5 Hacker Lifestyle

The act of accessing computer system and network without authorization is called hacking. The person who conducts this activity is referred to as hackers. The attacker will conduct many preattack activities in order to obtain information successfully. Frequently, they can alter or change or modify security systems for the success of hacker's business goal; it can differ from the actual purpose of the system [1]. Nowadays, many organizations hire hackers as their staff to find system flaws, weakness of organization, and vulnerable areas to be a good security system or organization and prevent malicious hackers who are breaking into the system [1].

2.5.1 White Hat Hacker

White hat hackers are good person and also referred to as an ethical hacker. They are working with organizations to harden the security system and to get more profit for the organization. A white hat hacker legally has permission to exploit the targeted system of the organization and compromise or tested machines within the prescribed rules or setting in advanced rules of engagement. Individual hacker specializes in and analyzed ethical hacking tools, techniques, and methodologies to secure an organization's information systems [1].

Not having the same function of the black hat hacker does, ethical hackers have to exploit security network flaws and look for backdoors when they are legally permitted to do so and get an authority. This kind of hackers always tries to disclose every vulnerability they find in the company's security system so that it can be fixed before they are being exploited by malicious actors [1].

2.5.2 Black Hat Hacker

The meaning of black hat originated and came from Western movies, where the bad guys wore black hats and the good guys wore white hats [1].

A black hat hacker is an individual person who attempts to gain unauthorized entry into a system or network to exploit the organization's information for malicious reasons. The black hat hacker does not have any permission or authority to compromise their targets. They try to inflict damage by compromising security systems, altering functions of websites and networks, or shutting down systems. They often do so to steal or gain access to passwords, financial information, and other personal data [1].

2.5.3 Gray Hat Hacker

Grey hat hacker exploits networks and computer systems in the way like a black hat hacker does. They do not intend to do any malicious activity like black hat hacker, disclosing all loopholes and vulnerabilities to law enforcement agencies or intelligence agencies. Usually, grey hat hackers hack into computer systems to notify the administrator or the owner that their system/network contains one or more vulnerabilities in order to be fixed immediately. Grey hats hacker may also extort the hacked, offering to correct the defect for a minimal cost [1].

2.6 Cyber-Attack Lifecycle

The characteristic of cyber-attack lifecycle can be categorized as follows:

2.6.1 Phase 1: Reconnaissance

This is the initial step of information gathering stage. Attacker collects targeted organization's information as possible as he can via several ways in order to satisfy the goal of the attackers such as financial gain, intentional brand damage, and accessing sensitive information [14].

2.6.2 Phase 2: Initial Compromise

After reconnaissance stage, the next step is initial compromise phase. In this stage, attackers break out the system, gaining access to the internal network via bypassing the defended perimeter (Firewall). Sometimes, hacker used by compromising user account or system. Attacker always used to deceive the victim with phishing email or spear phishing attack. Once the victim's machine gets an email in their mail inbox and attachment is opened, this suddenly generated malware is controlled by the attacker. This stage can get benefit for attacker to get further instructions such as lateral movement [14].

2.6.3 Phase 3: Establish Foothold

After phase 2 stage, this phase immediately followed the initial compromise step. Typically, this phase 3 involves the attacker downloading some malicious software in order to establish command and control and persistent, long-term, remote access to victim's machine [14].

2.6.4 Phase 4: Lateral Movement

After being connected and gaining access to the internal network, attacker uses lateral movement to look for another additional system to compromise user account, privilege level. Attackers use this technique to go further step through the network when they search for the organization's key data and assets [14].

2.6.5 Phase 5: Target Attainment

After the malicious software activities have established at a lateral movement connection with multiple machines in the network, another step will be carried out for unsolicited authorization, account compromising, and privilege escalation [14].

2.6.6 Phase 6: Ex-filtration, Corruption, and Disruption

This is the final phase of attack, the permission escalated are used to transmit data whole of the network, it is called ex-filtration [14]. They steal sensitive information from the business organization and corrupted critical information resources including delete file and disruption.

2.6.7 Phase 7: Malicious Activities (Fig. 2.10)

Normally, threat lifecycle in cloud computing is the same as the traditional network environment.

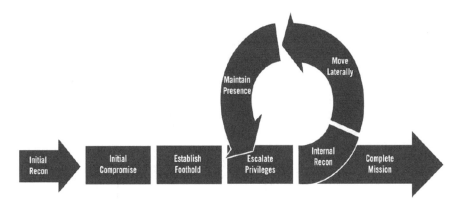

Fig. 2.10 Hacker attack life cycle

Table 2.1 IP addresses domain of experimental lab

No	FQDN	IP address	System	Installations
1	vcenter.domain1.site	172.16.10.1	Windows 2012 R2	DNS, DHCP, NAT, NTP, and VMware vCenter Server
2	esx1.domain1.site	172.16.10.11	VMware ESX	ESXi 5.0
3	esx2.domain1.site	172.16.10.12	VMware ESX	ESXi 5.0
4	nas1 .domam1, site	172.16.10.21	NAS	Openfiler
5	research.domain1. site	DHCP	Windows XP (Management PC)	VMware vSphere Client

2.7 Implementation Framework for Protecting Mechanism

2.7.1 VM Creation of Virtual Network Configuration

In this experiment, the system tested both type 1 and type 2 hypervisor attack on hypervisor hijacking attack.

As first step of proposed system design, a virtual environment was created for both type 1 and type 2, and the networks were configured on the host server machine, with one network allowing access to the Internet and an internal one with the IP address range shown in Table 2.1 and Fig. 2.14 as IP address domains. For type 2 hypervisor, 10.0.0.0/24 is used to communicate between the virtual machines. Then, the Management VM was created with Linux open-source software through a desktop environment installed, before the network interfaces were configured and SSH (Secure Socket Shell) access was enabled for remote access.

The next step is the creation of the basic VM template which was used to create a total of three VMs for the internal network by installing their respective servers one by one (Figs. 2.11 and 2.12).

Fig. 2.11 Configuration of virtual machine in experimental lab

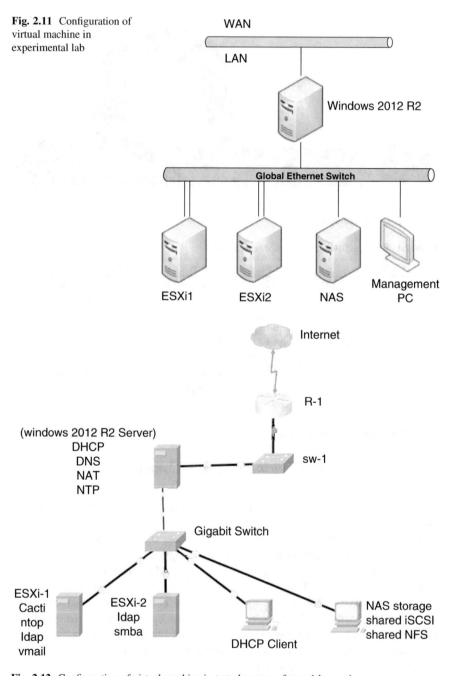

Fig. 2.12 Configuration of virtual machine in tested system of type 1 hypervisor

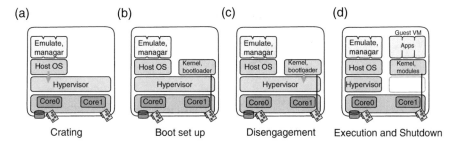

Fig. 2.13 Configuration bridging of virtual machine in type 1 hypervisor

During the setup, the following stages are configured:

- Free version of hypervisor to manage the virtual machine
- Required to navigate each hypervisor to monitor for attackers
- Configuring the pre-allocation of processor cores and memory resources
- Using virtualized I/O devices to apply services
- Monitoring modifications to the guest OS to perform all system discovery during boot setup
- Setting up (NIC) card for the guest virtual machine in more direct contact with the underlying hardware (Figs. 2.13, 2.14, 2.15, and 2.16)

2.7.2 Tested Methodology

In VM preparation for type 2

Step 1: Kali Linux is configured as main attacking platform.
Step 2: In virtual box configuration, change Kali network NIC for NAT mode to Bridge.
Step 3: Setting login password with username = root, password = root.
Step 4: Open terminal window and type ifconfig command.

The IP address of Kali Linux will show 10.0.0.12/24.
The system used the following steps to achieve successful VM configuration. After the package installation process, it can use the following commands.

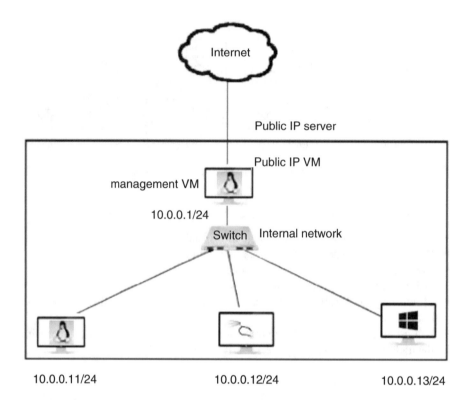

Fig. 2.14 Tested in type 1 hypervisor environment of virtual machine

```
[    0.000000] Initializing cgroup subsys cpu
[    0.000000] Initializing cgroup subsys cpuacct
[    0.000000] Linux version 3.13.0-24-generic (buildd@panlong) (gcc version 4.8.2 (Ubuntu 4.8.2-19u
-Ubuntu SMP Thu Apr 10 19:11:08 UTC 2014 (Ubuntu 3.13.0-24.46-generic 3.13.9)
[    0.000000] Command line: BOOT_IMAGE=/boot/vmlinuz-3.13.0-24-generic root=UUID=05088af6-6cef-4516
2ea0a ro
[    0.000000] KERNEL supported cpus:
[    0.000000]   Intel GenuineIntel
[    0.000000]   AMD AuthenticAMD
[    0.000000]   Centaur CentaurHauls
[    0.000000] e820: BIOS-provided physical RAM map:
[    0.000000] BIOS-e820: [mem 0x0000000000000000-0x000000000009fbff] usable
[    0.000000] BIOS-e820: [mem 0x000000000009fc00-0x000000000009ffff] reserved
[    0.000000] BIOS-e820: [mem 0x00000000000f0000-0x00000000000fffff] reserved
[    0.000000] BIOS-e820: [mem 0x0000000000100000-0x000000007fffdfff] usable
[    0.000000] BIOS-e820: [mem 0x000000007fffe000-0x000000007fffffff] reserved
[    0.000000] BIOS-e820: [mem 0x00000000feffc000-0x00000000feffffff] reserved
[    0.000000] BIOS-e820: [mem 0x00000000fffc0000-0x00000000ffffffff] reserved
[    0.000000] NX (Execute Disable) protection: active
[    0.000000] SMBIOS 2.4 present.
[    0.000000] DMI: QEMU Standard PC (i440FX + PIIX, 1996), BIOS Bochs 01/01/2011
[    0.000000] Hypervisor detected: KVM
[    0.000000] e820: update [mem 0x00000000-0x00000fff] usable ==> reserved
[    0.000000] e820: remove [mem 0x000a0000-0x000fffff] usable
[    0.000000] No AGP bridge found
```

Fig. 2.15 Command line tested to show hypervisor type

Fig. 2.16 Perl script of command line in detail

```
#!/bin/bash -
# virt-what.  Generated from virt-what.in by configure.
# Copyright (C) 2008-2011 Red Hat Inc.
# Do not allow unset variables, and set defaults.
set -u
root=''
skip_qemu_kvm=false

VERSION="1.13"

function fail {
    echo "virt-what: $1" >&2
    exit 1
}

function usage {
    echo "virt-what [options]"
    echo "Options:"
    echo "  --help        Display this help"
    echo "  --version     Display version and exit"
    exit 0
}

# Handle the command line arguments, if any.

TEMP=$(getopt -o v --long help --long version --long test-root: -n 'virt-wh
if [ $? != 0 ]; then exit 1; fi
eval set -- "$TEMP"

while true; do
    case "$1" in
        --help) usage ;;
        --test-root)
            # Deliberately undocumented: used for 'make check'.
            root="$2"
            shift 2
```

Step 1: #service libvirtd start
Step 2: #chkconfig libvirtd on
Step 3: #systemctl enable libvirtd.service
Step 4: #virsh list
Step 5: #mkdir -p /home/vmsserver-FM
Step 6: [root@server] # virsh
Step 7: virsh # help
Step 8: virsh # list –all
Step 9: virsh #dominfo server-VM
Step 10: virsh # start server-VM
Step 11: virsh # shutdown server-VM

Many virtualization implementations run on Linux system (some also can run windows). Some are quite easier to set up and manage than other kind of implementation method. In this section, the system tested both high-level virtualization and common virtualization concepts as a type 1 (bare metal) hypervisor and type 2 (hosted) hypervisor.

2.8   Behavior-Based Analysis for Hypervisor Detection

Intrusion detection systems (IDSs) have a vital role for infrastructure security in cloud computing which are designed to detect log and incident response for using

Fig. 2.17 Cloud-based
intrusion detection system

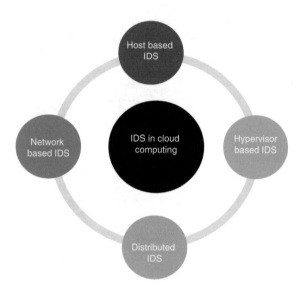

unauthorized behavior both in real time and after the event. In cloud computing, there are four types of IDSs systems such as network-based IDS, host-based IDS, hypervisor-based IDS, and distributed IDS as shown in Fig. 2.17.

Behavior-based detection approach is a proactive approach method to manage security incidents that include monitoring and see log process for abnormal behavior of users, devices, network, and servers in order to block (stop) suspicious events. This model depends on a collection set of normal behavior like users and traffics. If the activities or pattern is considered as abnormal, the system is regarded as suspicious (malicious) activities. Behavior-based detection approach analyzes data and application based on unknown attacks. On other side, knowledge-based technique is used known attack. For unknown attack, there can be high false positive rates; any new traffic pattern can be regarded as suspicious. This model can detect zero-day attack (Table 2.2).

2.9 Protecting and Mitigation Technique for System Hardening

The major important primary technique for secure hypervisor is to keep separate from usual traffic. The access to guest operating system should be defined by restricting the access to the hypervisors with the intention of avoiding hyper-jacking attacks (Tables 2.3, 2.4, and 2.5).

Table 2.2 Analysis of cloud-based detection system

Type of IDS	Features	Limitations/challenges	Position in cloud	Deployment and monitoring
Host-based IDS	Identify intrusions by monitoring system activity (Logs, Files system, etc.)	It must be installed on each machine (VM, hypervisor, host). Monitors the machine where it is deployed	In each virtual machine, in hypervisor, or in host machine	In VMs: cloud user. In hypervisor or in a host machine: cloud provider
Network-based IDS	Identify intrusions by monitoring network traffics	Difficulties of detecting intrusions in a virtual network. Can only detect intrusions coming from the network where it is deployed	In an external or virtual network	Cloud provider
Hypervisor-based IDS	Monitor and analyze communication between VMs, between hypervisor and VMs and between hypervisor and virtual network. It is the most suitable for cloud	Resources on the subject are limited. Difficult to understand	In hypervisor	Cloud provider
Distributed IDS	It allows a company to efficiently manage its incident analysis resources and to identify threats to the network across multiple network segments	The adoption of DIDS is still challenging due to the complex architecture of the infrastructure and the distinct kinds of users lead to different requirements and possibilities for being secured	It operates both on host and network	Cloud provider

Table 2.3 Analysis of hyper-jacking attack and mitigation technique

Attack mechanism	Mitigation technique
DoS Attack	Disable IP broadcast Disable unused services Deploy firewall rules Deploy IDS policy and rules Apply security patches on host
Live VM migration	Encryption of data by the hypervisor Use IPsec tunnel Source virtual machine monitor level virtual firewall Destination virtual machine monitor level virtual firewall
Hyper-jacking	Separate traffic from usual traffic Restrict the access Regular patches
VM escape	Should be provided access based on role Host should run only the required resource-sharing functionalities/services Guest OS should run less number of application to avoid any pen test
VM sprawl	Restrict the access Should be provided access based on role Should be made periodic verification Properly turn off identified idle VM
Guest OS vulnerabilities	Periodically perform the patching process Adopt firewall applications and traffic Really needed applications should be installed
Hyper-wall	Combination of another model for more security Confidentiality and integrity protection Multi-facilitated functions compared with others

Table 2.4 shows some integrity of virtual machine monitoring security techniques

Tools	Description
Hyper-safe	The lightweight approach of hypervisor for controlling flow of integrity check to provide lifetime
Hyper-sentry	To provide stealthy and in context measurement for precisely measuring integrity System management mode weakness and limitation, NOVA appears

Table 2.5 shows control flow integrity of virtual machine

Tools	Description for mitigation
Non-bypassable memory lockdown	Endue VMM with self-protection Protect memory page and attribute from malicious modification
Restricted pointer Indexing	Memory page and control data protection

2.10 Future Studies

Going forward for future research, this paper will discuss in more detail measurement of security that yields comparison results to take achievement in more virtualized implementations to meet the needs of current customer issues and requirements.

2.11 Conclusion

In this proposed system, a very recent and fresh comprehensive survey on virtualization threats and vulnerabilities is presented with the classification of hypervisor hijacking attacks with existing defense mechanisms. The purpose of this paper is introducing the approach that intended to harden and protect the business and organization's values in the virtualization environment which is mostly used in modern cloud architecture. The system tested both type 1 and type 2 hypervisors. Actually, there is a very rare report of type 1 hypervisor hacks, but according to the theoretical assumption, any cybercriminal can run a program and break down the system of virtual machine. So as an administrator's point of view, he/she should be prepared for this kind of attack as an offensive nature. So, hardening hypervisor hijacking attack on the virtual reality is an actual successful proof of concept as a real-world threat. Although virtualization threats and attacks are listed and categorized in this paper, luckily, high impact hypervisor attack on virtualization can be avoided by using behavior-based detecting method. Most of the motivation and methods of traditional threats and attacks are basically the same in VM environment. So, administrators may have to encounter and face with similar attack technique. Anyway, security is very long and hard process to protect the organization, for all layers to harden.

References

1. http://www.ukessays.com/
2. https://www.zdnet.com/article/detecting-the-blue-pill-hypervisor-rootkit-is-possible-but-not-trivial/
3. D. Morabito, Detecting hardware-assisted hypervisor rootkits within nested virtualized environments. Master's thesis, AFIT/GCO/ENG/12-20, WrightPatterson Air Force Base, OH, USA, Accession Number ADA563168, June 2012
4. J. Rutkowska, Subverting VistaTM kernel for fun and profit, http://www.blackhat.com/presentations/bh-usa-06/BH-US-06-Rutkowska.pdf
5. D.D. Zovi, Hardware virtualization rootkits, http://www.blackhat.com/presentations/bh-usa-06/BH-US-06-Zovi.pdf
6. D. Ruest, N. Ruest, *Virtualization: A Beginner's Guide* (McGrawHill Publication, New York, 2009), p. 25

7. I. Korkin, Two challenges of stealthy hypervisors detection: Time cheating and data fluctuations. J. Digit. Forensic Secur. Law
8. https://www.webopedia.com/TERM/O/operating_system_virtualization.html
9. https://searchservervirtualization.techtarget.com/definition/hardware-emulation
10. http://www.wikipedia.com
11. A.M. Azab, P. Ning, Z. Wang, X. Jiang, X. Zhang, N. C. Skalsky, HyperSentry: Enabling stealthy in-context measurement of hypervisor integrity. Proceedings of the 17th ACM Conference on Computer and Communications Security CCS, 2010, pp. 38–49
12. M. Ali, S.U. Khan, A.V. Vasilakos, Security in cloud computing: Opportunities and challenges. Inf. Sci. **305**, 357–383 (2015)
13. https://pentestlab.blog
14. https://www.helpnetsecurity.com/2017/03/06/cyber-attack-lifecycle/

Chapter 3
Proposed Effective Feature Extraction and Selection for Malicious Software Classification

Cho Cho San and Mie Mie Su Thwin

3.1 Introduction

Due to the growing of malicious code in the information technology and cyber security, the knowledge and understanding of new unknown malicious code or program protection is an important trend in the suspicious software detection system using machine learning (ML) methods. There are normally two ways to carry out the suspicious software analysis, static and dynamic analysis for detecting and finding the new malware.

The process of analyzing the software or program without executing the program is referred to static analysis that can classify and detect the known and unknown malicious code [1]. It is the first approach for analyzing and detecting the malicious software that has been stated in [2]. The static malware analysis examines the assembly code of binary file to identify the retrieve flow of code and sequential instructions without actually executing the executable sample [7]. Reverse engineering is a common approach to extract static information from a binary. Disassembly and hexadecimal dumping of binary file are the two main techniques to pre-process and get static features from the sample [4]. A disassembler tool can be used to decompile Windows executable files, such as IDA Pro and OllyDbg, that display assembly instructions, provide information about the malware, and extract patterns to identify the attacker' desire.

In [5] static or code analysis is faster and simple than the other analysis. However, it cannot be effective for obfuscated and complex malware and might leave the significant malicious behaviors. Additionally, the obfuscation techniques, polymorphism, metamorphism, compression, encryption, and run-time packing,

C. C. San (✉) · M. M. S. Thwin
Cyber Security Research Lab, University of Computer Studies, Yangon, Myanmar
e-mail: chochosan@ucsy.edu.mm; drmiemiesuthwin@ucsy.edu.mm

© Springer Nature Switzerland AG 2019
G.R. Sinha (ed.), *Advances in Biometrics*,
https://doi.org/10.1007/978-3-030-30436-2_3

introduced by malware authors, lead static analysis complicated, time-consuming, and nearly unfeasible. Thus, malware analysts and researchers developed and performed dynamic method which is more effective than static to the obfuscation techniques.

The dynamic or run-time analysis method performs the running or executing the malware in a safe and controlled environment. It inspects the malware dynamic behavior to decide which function or system calls are intercepted sequentially, which is also called hooking method, by a malware to determine the nature of suspicious file behavior in a virtual machine environment [7].

Another way to analyze the malicious file is using sandbox. NIST defines the sandbox in [8]: it is a security model where applications/programs are run within a safe environment or a sandbox. Sandboxes record the changes of file system, registry keys, and network traffic and then generate standardized report format. There are common sandboxes that can leverage a quick analysis for malicious files. Sandboxes such as GFI Sandbox, Anubis, Joe Sandbox, ThreatExpert, and Cuckoo Sandbox can analyze malware for free. Cuckoo Sandbox has been used to discover the malware behavior patterns in our approach.

Figure 3.1 shows different types of static and dynamic features used for malware detection and classification in recent researches.

According to [4] dynamic features can be derived from host trace and network trace-based features. The activities of internal memory, files and file system, registry, hardware performance counters, and status of running processes of host are considered as host trace features. The proposed system used the API n-gram (where $n = 1,2,3$), from host trace of dynamic analysis. API is the most common used attributes or features in malware analysis. The proposed system used the Cuckoo Sandbox for analyzing the samples. The APISTATS from JSON report of sandbox

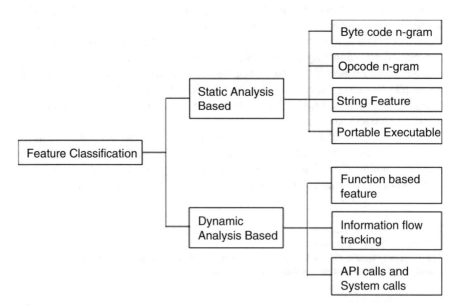

Fig. 3.1 Features based on types of analysis

are extracted as feature APIs for classification system. The number of extracted API from APISTATS is 306 for malware and benign. The malware samples contain six different categories such as Adware, Backdoor, Downloader, Trojan, Virus, and Worm.

The proposed system applies the n-gram technique as the extracted attributes or features depend on each other. However, bigram will only be applied because the system concerns the processing/run time of the classifier. And the experiment shows that these grams are enough to distinguish malware from benign. The proposed system applies two machine learning (ML) techniques and two feature selection (FS) approaches to differentiate the malware from benign. The sklearn ML library [6] has been used to apply these ML techniques and FS approaches. The next section will describe the related works that perform the classification, detection, and feature selection through dynamic analysis.

The content of the paper is structured into five sections. The recent related research work is described in the next section, and the proposed feature extraction, selection, and classification system are provided in Sect. 3.3. Section 3.4 supports the experiment and results discussion, respectively. Last section highlights the conclusion and the research plans for future.

3.2 Related Work

This section presents the current research works that have done through dynamic or run-time analysis. Researchers are now working by proposing the hybrid nature on both analysis and features such as hybrid analysis and hybrid features combination and feature fusion methodologies. And most of the attributes that use to detect and identify the malicious programs are API which is based on the number of occurrences of API (frequency), the order of API (sequence), and system calls.

The dynamic analysis means the malicious behavior is monitored by tracing or inspecting the API calls from Windows and network connections by running the suspicious files in a safe environment. The extracted API function calls are used to detect malicious behavior through behavior or dynamic or run-time analysis method. The API calls from different categories such as process, registry, file, and network contain the function names, return values, and parameters of an executable. The dynamic analysis extracts distinct features to find the malicious software using the API sequence and frequency [5, 11]. The frequency on API might indicate how API calls play an important role for a malicious file, while API sequence shows the knowledge about how important consecutive behaviors of the malware are. Moreover, the researchers also utilized additional behaviors as features beside API calls which are dynamic link library (DLL), file opened/closed, and mutex that provide useful data about the suspicious files [31].

Most of the dynamic techniques focused on API calls [9–12] to represent malware behaviors. The authors used TEMU for dynamic analysis module which is based on QEMU. They collected the API calls and other essential information of running malware and then established the multilayer dependency chain [9].

The authors in [14] performed classification through run-time analysis using Cuckoo. The total number of samples, 42,068, was used for classification, 67% was used for training set, and 33% was used for the testing set. The authors extracted and used 151 API calls as main features; the first 200 API calls were used for sequence. They combined the features of 24 API FBs, modified sequence of first 40 different API, and 4 counters captured by modifying the Cuckoo Sandbox. In their approach, they employed a combination of features that achieved an average weighted AUC value of 0.98, TPR of 0.896, and FPR of 0.049 by applying RF classifier.

In [9], the authors proposed the variants of malware classification technique based on behavior profile. The authors used the TEMU to monitor the malware behaviors. They captured the API calls and other information and then established multilayer dependency chain by converting the function flow into multilayer behavior chain. To assess the validity and accuracy of the method, they downloaded 200 samples of 12 types from Anubis website. To identify the malware variants similarity, similarity comparison algorithm had been used in their work.

In [15], the authors experimented a 552 PE dataset with their corresponding API calls. These samples were executed in a Windows 7 virtual environment using Cuckoo Sandbox. Tf-idf (term frequency-inverse document frequency) had been used to extract relevant 4-gram API call features. The authors used four machine learning methods for training and testing the data. They got the accuracy between 92% and 96.4%. In [16], 2 malware datasets had been created such as 10 families and 10 different types. Then the authors extracted the features by using the memory access patterns recording technique from the sample. Then the authors performed n-grams size of 96. N-grams apply on the features of dynamic and static.

In [17], the authors extracted separately different features through run-time analysis, likewise API call, the usage of system library, and the operations. Four different classifiers and correlation-based feature selection method from WEKA tool had been applied in their work. Bigram API and API frequency approaches give the best performance by using the RF for four datasets. In [13], the authors conducted the detection and classification system using the calls of API sequences for four different families including normal group.

Masud et al. used information gain after the n-grams extracting to select the best 500 features. They experimented on two different datasets: first dataset contains 1435 executable files (597 cleanware and 838 malware), and the second dataset contains 2452 executables (1370 clean and 1082 malware). The information gain (IG) attribute selection method was used in [18–20] and their accuracies with 98%, 94.6%, and 97.7%, respectively. The accuracy of the hybrid model was 97.4 for both datasets $n = 6, 4$, respectively [21]. The chi-square feature selection was applied in [22, 23], and Tf-idf was applied in [24] to get the most relevant features.

The proposed system has also used the Chi2 χ^2 and PCA methods to support the high performance for classification by reducing the features size. The proposed approach provides the over 99% of accuracy on 300 features using χ^2 feature selection approach and 10 features with PCA approach on unigram. This section presents the existing works related to malware classification using machine learning algorithms and supports previous researches about feature extraction methods based on dynamic malware analysis and classification. These research efforts use different malware modeling techniques using static or dynamic features obtained from malware samples.

3.3 Malicious Software Family Classification System

Malicious software classification is not a new topic but it is still needing attention and solution to be solved for cyber threats nowadays. Many researches have been carried out to analyze and classify the malicious files using the API function calls sequences to model malicious behavior. Thus, the malware behavioral patterns can be obtained by understanding the API Call Sequences (API-CS). Therefore, the proposed system also used the API-CS by proposing the API Feature Extraction Procedure and applying the n-gram method. Figure 3.2 shows the step-by-step process of malicious software analysis architecture for the proposed system.

Fig. 3.2 Malicious software analysis and classification system

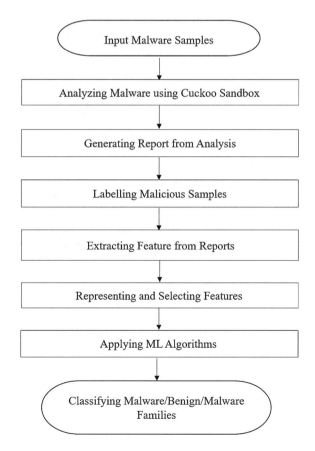

Malicious samples have been collected from virus share[1]; the proposed system experimented nearly 25,000 from 6 different families. However, the proposed system discards the samples based on the following conditions:

1. If the analysis report does not contain Virus Total (VT) label results
2. If the family does not have at least 1500 samples
3. If the extracted API features from report do not have at least 15 API features without duplicate ones

Therefore, the experiment provided a total of 20,809 samples from 6 malware families and cleanware in this research work. And Table 3.1 describes the number of samples for 14 different families and target label (Class) for each family. Table 3.2 describes the six different families for malware class.

3.3.1 Analyzing Malware Samples and Generating Reports

Cuckoo Sandbox [32] has been used to perform the dynamic analysis in this system. Windows 7 Operating System (OS) has been used for the virtual environment for analysis in Virtual Box and Ubuntu as host OS. It is widely used and open source for the researchers of academic and independent from a small to large business enterprises. It can analyze different types of malicious files, such as executables files, office document files, PDF files, emails, etc., and malicious websites. And it can also trace the API calls and the behavior of file, and dump and analyze network traffic, even encrypted with an SSL/TLS.

It generates the reports from analysis with multiple formats such as HTML, JSON, and PDF formats. But the proposed system used JSON format to extract the malicious behaviors. Figure 3.3 shows the lab setup environment of malware analysis. The lab setup environment has been described in the following figure. Ubuntu 18.4 (host) OS and Windows 7 (guest) have been used for analyzing the malicious samples on sandbox. The normal applications such as Office Documents, Adobe Reader, Browsers, etc. have been installed on virtual OS.

Table 3.1 Malware/benign dataset for experiment

Name	# of samples	Target label (Class)
Clean	5420	0
Malicious	15,389	1
Total number of samples	**20,809**	**2 (0/1)**

---

[1] http://tracker.virusshare.com:6969/

Table 3.2 Six different categories of malware

Categories	# of samples
Adware	2200
Backdoor	2076
Virus	1740
Trojan	5000
Worm	2397
Downloader	1976
Total	**15,389**

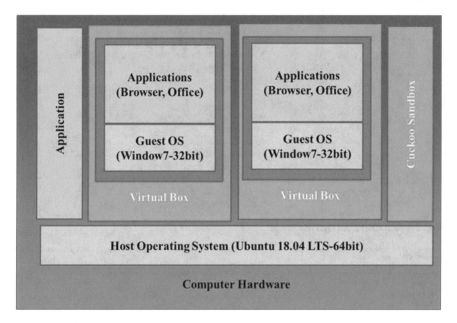

Fig. 3.3 Lab setup environment for malware analysis

3.3.2 Labeling Malicious Samples

The report of sandbox provides the VT label for each malicious sample if the analyzed sample exists on the VT database. Figure 3.4 describes the VT scans results with their anti-virus vendors, respectively.

The proposed system extracted these results using the regular expression (RE) theory. RE is a very powerful, useful, efficient, and flexible text processing language. And then count the occurrence of each results that extracted from RE, and choose the maximum value of word as label for each sample. Figure 3.5 shows the example of choosing the label for each malicious sample. These labels indicate the single malicious file. According to Fig. 3.5, this kind of malware sample will be labelled as Adware, and it is an Adware category or family.

```
721    "virustotal": {
722        "scans": {
723            "Bkav": {
724                "detected": true,
725                "version": "1.3.0.9899",
726                "result": "W32.HfsAdware.88A5",
727                "normalized": [
728                    "HfsAdware"
729                ],
730                "update": "20190130"
731            },
732            "MicroWorld-eScan": {
733                "detected": true,
734                "version": "14.0.297.0",
735                "result": "Adware.Generic.1775594",
736                "normalized": [],
737                "update": "20190131"
738            },
739            "CMC": {
740                "detected": false,
741                "version": "1.1.0.977",
742                "result": null,
743                "normalized": [],
744                "update": "20190131"
745            },
746            "CAT-QuickHeal": {
747                "detected": true,
```

Fig. 3.4 VT label from analysis report

```
38,adware
32,adposhel
31,trojan
18,gen
13,generickd
12,pua
9,razy
8,generic
8,malicious
7,variant
6,adw
5,riskware
5,virus
```

Fig. 3.5 Choosing label for each malicious sample

3.3.3   Extracting Malicious Features

After categorizing the malicious family, the process of extracting the malicious features has been performed in the proposed system. The APISTATS result has been extracted from the JSON for API features. And the procedure of extracting APISTATS process is described as followed:

```
APISTATS Feature Extraction Procedure
Input: JSON reports
Output: extracted API files fᵢ
1: begin
2:              if (JSON ≤ JSONs)
3:                 try
4:                      data = []
5:                   data = json.load(JSON)
6:                      try
7:                         api = data['behavior'] ['apistats']
8:                 print (api)
9:                 except KeyError:
10:                       print ("APIStats KeyError")
11:              except ValueError:
12:                    print ("JSONDecodeError")
13:              end if
14: end
```

The extracted raw APIs attributes are extremely large, and it is not able to handle the classification system. So, data cleaning processes have been provided in this phase of proposed system. The raw data cleansing processes are described as followed:

1. Remove empty line if the extracted API files have empty line
2. Remove noise data such as comma, colon, single code, double code, curly braces, and so on.
3. Remove duplicate API by keeping the order of API calls
4. Discard the extracted API files if the number of APIs does not have at least 15 API.

3.3.4 Applying N-gram

After processing the data cleansing steps that are described above, n-gram method has been applied to ensure the identifying of malicious files. It is a continuous sequence of nth items from a given sequence. It is very useful for characterizing the sequences in natural language processing and DNA sequencing areas [3]. It has been adopted to extract the sequence of features in malware classification for static and dynamic analysis. But the static is the one mostly used n-gram such as opcode n-gram, byte-code n-gram, and API call n-grams. The proposed system applies the n-gram technique, where $n = 1,2,3$, to identify the malicious families and benign. The total number of APIs after processing the data cleansing stage is 306 APIs. Therefore, the number of features for unigram (1G) is 306. Then, the number of features for bigrams (2G) is 10,796 g. And the total number of samples for classification is 20,809 instances. The explosive number of features will increase as long as the n number increases in dataset. Moreover, it could lead to an overfitting. So the proposed system used the unigram and bigrams for the classification. Both unigram and bigram provide a high accuracy to distinguish malware and benign.

3.3.5 Representing and Selecting Malicious Features

Attributes representation process has been conducted after applying n-grams on extracted APIs. The process of attributes representation has been performed based on the presence and absence of features in global feature database. The proposed system used the binary feature vector representation that is described in our previous research work [25] and described as follows:

$$\text{API}_i = \begin{cases} 1, & \text{if API is in MBAPIDS File} \\ 0, & \text{otherwise} \end{cases}$$

The total global database MBAPIDB (Malware Benign API Database) contains all API features of malware and benign. If the extracted API contains in MBAPIDB, it is denoted as 1, and if not, it is denoted as 0. For example, the sample F1 is the single malware instance feature representation, and the last item 1 is the class label for malware family.

$$\text{F1} = \{1, 1, 1, 0, 1, 0, 1, 0, 0, 1, \ldots, 1\}$$

The next step is the selection of feature for classification. The purposes of applying the selection approaches are to select the appropriate features to the target class and to minimize the processing time. Feature or attribute selection methods are used for reducing the size of a feature dataset. The key role of feature selection process is to improve the classification performance as well as improving the detection accuracy by choosing or transforming the feature set. Subsequently, the processing time for classification process can speed up and improve the evaluation results since the feature number is reduced.

Among the three FS methods such as filter, wrapper, and embedded methods, most researchers commonly used the filter-based approach in malicious classification and detection research areas. The filtering approach does not depend on any particular algorithm. It is very fast and computationally less expensive than the other two methods. It is easy to scale to very high-dimension datasets [30]. So, the proposed system applies the χ^2 method from filtering approach.

The proposed system used the two feature selection methods from sklearn. The efficiency of classification system can be improved by applying the attribute selection techniques such as chi-square (χ^2) and principal component analysis (PCA).

Chi-Square (χ^2) It is a statistic approach and very effective for feature selection process. The proposed system chooses χ^2 to select the feature because it can handle the multi-class data with an excellent performance. The proposed system used the implementation of χ^2 from sklearn ML library with python.

Principal Component Analysis (PCA) It is used to visualize and explore high-dimensional datasets. It reduces a set of possibly -correlated, high-dimensional variables to a lower-dimensional set of linearly uncorrelated synthetic variables called principal components. PCA reduces the dimensions of a dataset by projecting the data onto a lower-dimensional subspace [27].

3.3.6 Classifying Malware vs Benign Using Machine Learning

ML has powerful ability and capability to do many things for cybersecurity. It can be used to identify the advanced persistent threats (APTs) and zero-day attacks which are more complex than the normal malware or threats. And it can be used in many intrusion detection systems (IDS) because it can detect new and unknown attacks. It can be applied in many areas of information security such as spam and phishing email detection, phishing website detection, and virus detection. To classify the malicious and benign software, the proposed system used the two ML methods, random forest (RF) and K-nearest neighbor (KNN) from sklearn ML library.

Random Forest It combined the multiple decision trees, so it became an ensemble. It can handle the binary, categorical, continuous, and missing values, so it is suitable for high- dimensional data modeling. It can overcome the overfitting problems due to the nature of bootstrapping and ensemble scheme. Thus, it does no need to prune the trees. Besides high prediction accuracy, it is efficient, interpretable, and non-parametric for various types of datasets [28].

K-Nearest Neighbors (NN) It is an instance-based learning and also known as lazy learner. The lazy is called not because of its apparent simplicity, but because it doesn't learn a discriminative function from the training data but memorizes the training dataset instead. It is a sub-category of non-parametric approach [29].

The performance of ML classifiers has been evaluated using confusion matrix (CM), accuracy (ACC), precision recall (PR), and receiver operator characteristics area (ROC).

Confusion Matrix (CM) It is a popular way to describe a classification model. CM can be formed for binary and multi-class classification models. It has been created by comparing the predicted class label of a data point with its actual class label. After comparing the whole dataset repeatedly, the comparison results are formatted in a matrix form. This resultant matrix is the confusion matrix [26]. And Fig. 3.6 describes the typical structure of a CM.

Accuracy (ACC) It is a common evaluation method of a classifier performance. It is used to define as the percentage of overall accuracy of correct predictions. It can be calculated from the formula [26]:

Fig. 3.6 Confusion matrix
(CM) structure

		Predict	
		P	N
Actual	P	True Positive (TP)	False Negative (FN)
	N	False Positive (FP)	True Negative (TN)

$$\text{Accuracy} = \frac{TP + TN}{TP + FP + TN + FN} \tag{3.1}$$

Precision (P) It is a positive predictive value that can be achieved from CM. It is defined as the number of predictions made that is actually correct or relevant out of all the predictions based on the positive class [26]. It can be calculated from the following formula:

$$\text{Precision} = \frac{TP}{TP + FP} \tag{3.2}$$

Recall (R) It is also known as sensitivity, and it is used to identify the relevant data points with percentage. It is defined as the number of instances of the positive class that were correctly predicted [26]. It is also called as hit rate, coverage, or sensitivity. The value of recall can be computed as follows:

$$\text{Recall} = \frac{TP}{TP + FN} \tag{3.3}$$

Receiver Operating Characteristic (ROC) It can be used for both binary and multiclass classifiers. TP rate and the FP rate of a classifier are used to plot the ROC curve. TPR is also called recall or sensitivity, and it is the total number of correct positive results, predicted among all the positive samples in dataset. FPR is also defined as 1- specificity or false alarms, determining the total number of incorrect positive predictions among all negative samples in the dataset [26].

3.4 Results and Discussion

The total 25,000 malicious samples have been analyzed in this research work, but 20,809 benign samples have been used for family classification. The proposed system contributes the API feature extraction for malicious family classification.

After processing the raw data cleansing on extracted APIs, the remaining API features are 306 APIs. As the malicious behavioral patterns sometimes depend on the sequence of function calls and system calls, the proposed system applies the n-gram technique to ensure the right family classification. Therefore, the proposed system noted these APIs as unigram (1G). The total number of bigram (2G) API features is 10,796 g, and trigram (3G) features are 37,919 g. In this case, the trigram features are quite large, so the proposed system only considers to perform the unigram and bigram on our dataset since we concern the processing/run time. Table 3.3 shows the number of API grams for classification. After utilizing the n-grams on extracted APIs, the proposed system performs the attribute selection process before classifying families. The proposed system uses the two attribute selection methods, chi-square with SelectKBest and PCA, from sklearn. RF and kNN have been used to classify the malicious families and benign. The proposed system uses the 25% (5203 executables) for testing and the rest 75% (15,606 executables) for training. The proposed system uses accuracy, precision, recall, and ROC scores to assess the efficiency and effectiveness of extracted prominent APIs.

Table 3.4 describes the comparison of accuracy on unigram API (306 g) dataset using two FS and ML techniques. The proposed system compares the accuracies by selecting the five different numbers of features from unigram such as 10, 50, 100, 200, and 300 APIs. The RF classifier provides better accuracy 99% on selected 10 API using PCA and 300 API using χ^2 (Chi2). The kNN classifier provides 97% on χ^2 with 300 API and PCA with 10 API. The finding from the experiment is that the accuracy is increased when PCA chooses the small number of API. It is inversely proportional to the Chi2 approach. In χ^2, the accuracy has been increased as long as the selected API number is increased. RF classifier produces better accuracy on PCA with 10 features and χ^2 with 300 features than the kNN.

Figure 3.7 provides the confusion matrix results for RF on Chi2 χ^2 (300 API) and PCA (10 API). The correctly classified instances number of PCA on malware is slightly better than the Chi2's result.

Figures 3.8 and 3.9 describe the precision-recall (PR) curves and ROC curves of Chi2 for 300 API on RF and kNN classifiers.

Table 3.5 shows the comparison tables of accuracy on bigram API features. For bigram API selection, the proposed system used the different number of features

Table 3.3 Total number of grams after applying n-gram on extracted API dataset

N-grams	# of grams or APIs
Unigram	306
Bigrams	10,796

Table 3.4 Accuracy (%) comparison on selected unigram API

	RF					kNN				
	10	50	100	200	300	10	50	100	200	300
Chi2	86.2	97.2	97.9	98.5	**98.7**	82.9	95.9	96.4	96.4	**97**
PCA	**99**	98.7	98.7	98.7	98.7	**97.3**	97.2	97.2	97.2	97.1

Fig. 3.7 Confusion matrix of
RF classifier on unigram
dataset. (**a**) CM for RF on
selected 300 API using Chi2.
(**b**) CM for RF on selected 10
API using PCA

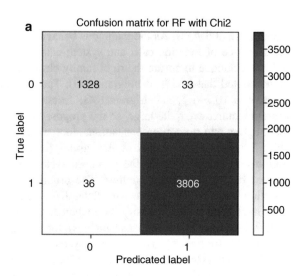

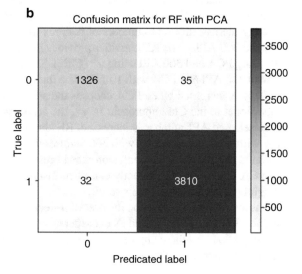

such as 100, 200, 300, 400, and 500, unlike unigram. Unigram has been selected
according to 10, 50, 100, 200, and 300.

The total number of bigram API is 10,796, and testing dataset is 5203 from
20,809 instances or samples. The training and testing dataset are split 75% and
25% of the dataset. The experiment shows that PCA increases the accuracy slightly
better than the Chi2 on both classifiers. Figure 3.10 and 3.11 depict the ROC and
PR curves for RF classifier using Chi2 and PCA for 500 g API.

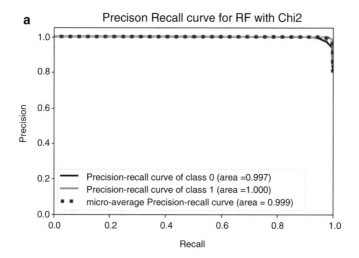

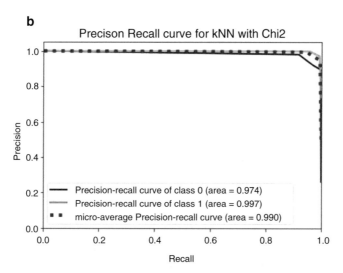

Fig. 3.8 PR curves for RF and kNN using Chi2 (χ^2). (**a**) PR curve of RF classifier on selected 300 APIs. (**b**) PR curve of kNN classifier on selected 300 APIs

Figure 3.12 shows the confusion matrix of RF classifier on selected 500 API bigram dataset using Chi2 (χ^2) and PCA. The confusion matrix results from PCA provide better than the Chi2 (χ^2) method, and the incorrectly classified instances are smaller than the Chi2 (χ^2).

Table 3.6 provides the accuracy comparison between our approach and other related works. Although the related work [22] is slightly better than our approach,

Fig. 3.9 ROC curves for RF and kNN using Chi2 (χ^2). (**a**) ROC curve of RF classifier on selected 300 APIs. (**b**) ROC curve of kNN classifier on selected 300 APIs

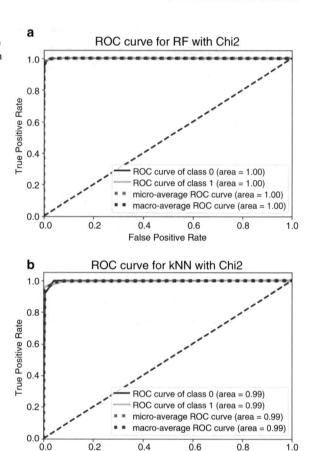

Table 3.5 Accuracy (%) comparison on selected bigram API

	RF					kNN				
	100	200	300	400	500	100	200	300	400	500
Chi2	95.5	97.2	97.9	98	**98.4**	94.5	95.8	96.1	96.7	**97.1**
PCA	98.8	99	99	99	**99**	97.9	98	98.1	98.1	**98.1**

the number of tested samples is quite small on both clean and malware. The original extracted API features provide the best accuracy of 99% on malware vs benign classification system. However, the proposed system applies the n-gram technique on extracted dataset since the proposed system concerns the malware that used the garbage code inserting techniques.

Fig. 3.10 ROC curves for RF classifier using Chi2 (χ^2) and PCA. (**a**) ROC curve on selected 500 g API using χ^2. (**b**) ROC curve on selected 500 g API using PCA

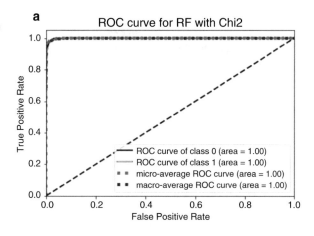

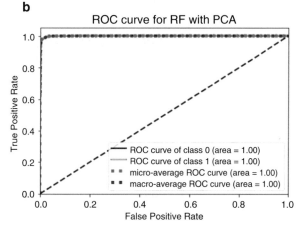

3.5 Conclusion

The usage of ML techniques in cyber security is becoming increasingly than ever before. The proposed system used the two ML methods to classify the malware vs benign for classification system. The proposed system contributes the malicious feature extraction for API features with n-gram and classification through dynamic analysis. The system extracts the API by using the APISTATS keyword from JSON report format. The proposed system has performed the raw data cleansing process after extracting the API from JOSN. Malicious JSON reports contain six different types of malware categories like Adware, Downloader, Trojan, Backdoor, Worm, and Virus. Two feature selection approaches, Chi2 χ^2 and PCA, have been conducted to reduce the size of features especially for bigram APIs as the size of n-gram feature is large to handle the classification. The results from the experiment

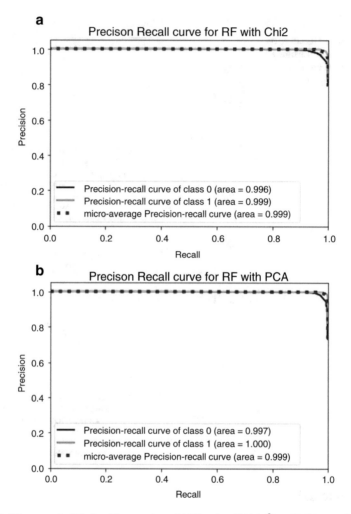

Fig. 3.11 PR curves for RF classifier on selected APIs using Chi2 (χ^2) and PCA. (**a**) PR curve of RF classifier on selected 500 APIs using χ^2. (**b**) PR curve of RF classifier on selected 500 APIs using PCA

can be noted that PCA provides better accuracy than the Chi2 χ^2 on unigram and bigram dataset.

In unigram, the accuracies of PCA are remained stable on different number of selected features. However, it is inversely proportional to the Chi2 approach. In Chi2 χ^2, the accuracy has been increased as long as the selected API number is increased. In bigram dataset, the accuracies of PCA also remain stable on both RF and kNN classifiers, while the accuracies of χ^2 vary on both RF and kNN classifiers. The proposed prominent feature extraction procedure provides a high accuracy with 99% and low FP and FN rates. The proposed system evaluates the performance of

Fig. 3.12 CM for RF classifier on 500 API. (**a**) CM for RF classifier on selected 500 API using Chi2 (χ^2). (**b**) CM for RF classifier on selected 500 API using PCA

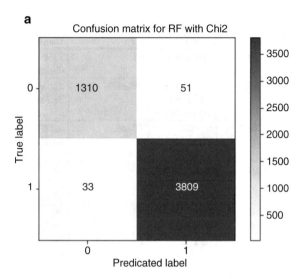

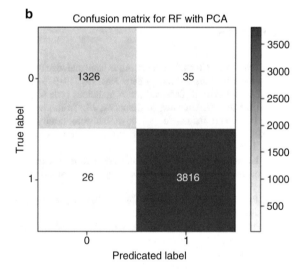

the classifier by using Accuracy, Precision-recall, and ROC scores. Moreover, the system also provides the low FP and FN rates on malware and benign classification.

The system will extend by adding the other malicious behavior features such as system library, process, and file opened/closed besides API in the future work. Moreover, the malicious samples from different families such as Zbot, Swizzor, Startpage, etc. will also be used to classify their families.

Table 3.6 Accuracy comparison for different FS approaches

Study	Features	FS	Dataset	Accuracy
[19] 2011	API string	IG	1368 malware 456 benign	97.4% for MFC 94.6% for M vs C
[31] 2012	API from 6 DLLs, input arguments	ReliefF	826 malicious 385 benign	98.4%
[24] 2015	n-gram (Static + Dynamic)	Tf-Idf	4288 samples from 9 families	~96%
[23] 2016	API, TPF	χ^2	338 malwares 214 benign 12 categories	98%
[22] 2016	Static + Dynamic	χ^2	7630 malware 1818 benign	99.60%
Our approach	API	χ^2 (unigram) PCA (bigram)	15,389 malicious 5420 benign 6 categories	99%

References

1. G. Tahan, L. Rokach, Y. Shahar, Automatic malware detection using common segment analysis and meta-features. J. Mach. Learn. Res. **13**, 949–979 (2012)
2. R. Lo, K. Levitt, R. Olsson, Mcf: A malicious code filter. Comput. Secur. **14**, 541–566 (1995)
3. M. Ahmadi, D. Ulyanov, S. Semenov, M. Trofimov, G. Giacinto, Novel feature extraction, selection and fusion for effective malware family classification. In Proceedings of the sixth ACM conference on data and application security and privacy, ACM, March, 2016, pp. 183–194
4. A. Kumar, A framework for malware detection with static features using machine learning algorithms, Doctoral dissertation, Department of Computer Science, Pondicherry University, 2017
5. A. Pektaş, Behavior based malware classification using online machine learning, Doctoral dissertation, Grenoble Alpes, 2015
6. F. Pedregosa, G. Varoquaux, A. Gramfort, V. Michel, B. Thirion, O. Grisel, J. Vanderplas, Scikit-learn: Machine learning in python. J. Mach. Learn. Res. **12**, 2825–2830 (2011)
7. J.Y.C. Cheng, T.S. Tsai, C.S. Yang, An information retrieval approach for malware classification based on windows API calls. 2013 International Conference on Machine Learning and Cybernetics, vol. 4, IEEE, 2013, July, pp. 1678–1683
8. M. Souppaya, K. Scarfone, *Guide to Malware Incident Prevention and Handling for Desktops and Laptops* (National Institute of Standards and Technology, Gaithersburg, 2013)
9. G. Liang, J. Pang, C. Dai, A behavior-based malware variant classification technique. Int. J. Inf. Educ. Technol. **6**, 291–295 (2016)
10. H.S. Galal, Y.B. Mahdy, M.A. Atiea, Behavior-based features model for malware detection. J. Comput. Virol. Hacking Tech. **12**(2), 59–67 (2016)
11. Y. Ki, E. Kim, H.K. Kim, A novel approach to detect malware based on API call sequence analysis. Int. J. Distrib. Sens. Netw. **2015**(6: 659101), 1–9 (2015)
12. C.-I. Fan, H.-W. Hsiao, C.-H. Chou, Y.-F. Tseng, Malware detection systems based on API log data mining. 2015 IEEE 39th Annual Computer Software and Applications Conference, 2015, pp. 255–260

13. M.A. Jerlin, K. Marimuthu, A new malware detection system using machine learning techniques for API call sequences. J. Appl. Secur. Res. **13**(1), 45–62 (2018)
14. R.S. Pirscoveanu, S.S. Hansen, T.M. Larsen, M. Stevanovic, J.M. Pedersen, A. Czech, Analysis of malware behavior: Type classification using machine learning. 2015 International Conference on Cyber Situational Awareness, Data Analytics and Assessment (CyberSA), IEEE, 2015, June, pp. 1–7
15. A. Ninyesiga, J. Ngubiri, Malware classification using API system calls. Int. J. Technol. Manag. **3**(2), 9–9 (2018)
16. S. Banin, G.O. Dyrkolbotn, Multinomial malware classification via low-level features. Digit. Investig. **26**, S107–S117 (2018)
17. A.G. Kakisim, M. Nar, N. Carkaci, I. Sogukpinar, Analysis and evaluation of dynamic feature-based malware detection methods, in *International Conference on Security for Information Technology and Communications*, (Springer, Cham, 2018), pp. 247–258
18. D. Komashinskiy, I. Kotenko, Malware detection by data mining techniques based on positionally dependent features. 2010 18th Euromicro conference on parallel, distributed and network-based processing, IEEE, 2010, pp. 617–623
19. V. Moonsamy, R. Tian, L. Batten, Feature reduction to speed up malware classification, in *Nordic Conference on Secure IT Systems*, (Springer, Berlin, Heidelberg, 2011), pp. 176–188
20. M. Mays, N. Drabinsky, S. Brandle, Feature selection for malware classification, In MAICS, 2017, pp. 165–170
21. M.M. Masud, L. Khan, B. Thuraisingham, A hybrid model to detect malicious executables. 2007 IEEE International Conference on Communications, IEEE, 2007, pp. 1443–1448
22. C. Cepeda, D.L.C. Tien, P. Ordóñez, Feature selection and improving classification performance for malware detection. 2016 IEEE International Conferences on Big Data and Cloud Computing (BDCloud), Social Computing and Networking (SocialCom), Sustainable Computing and Communications (SustainCom)(BDCloud-SocialCom-SustainCom), IEEE, 2016, pp. 560–566
23. M. Belaoued, S. Mazouzi, A chi-square-based decision for real-time malware detection using PE-file features. J. Inf. Process. Syst. **12**(4) (2016)
24. C.T. Lin, N.J. Wang, H. Xiao, C. Eckert, Feature selection and extraction for malware classification. J. Inf. Sci. Eng. **31**(3), 965–992 (2015)
25. C.C. San, M.M.S. Thwin, N.L. Htun, Malicious software family classification using machine learning multi-class classifiers, in *Computational Science and Technology: 5th ICCST 2018, Lecture Notes in Electrical Engineering*, vol. 481, (Springer, Singapore, 2018), pp. 423–433
26. D. Sarkar, R. Bali, T. Sharma, *Practical Machine Learning with Python: A Problem-Solver's Guide to Building Real-World Intelligent Systems*, 1st edn. (Apress, Berkely, 2017)
27. G. Hackeling, *Mastering Machine Learning with Scikit-Learn* (Packt Publishing Ltd, Birmingham B3 2PB, UK, 2017)
28. Y. Qi, Random forest for bioinformatics, http://www.cs.cmu.edu/
29. S. Raschka, *Python Machine Learning* (Packt Publishing Ltd, Birmingham B3 2PB, UK, 2015)
30. M. Khan, S.M.K. Quadri, Effects of using filter based feature selection on the performance of machine learners using different datasets. Bharati vidyapeeth's institute of computer applications and management's International Journal of Information Technology **5** (2013)
31. Z. Salehi, M. Ghiasi, A. Sami, A miner for malware detection based on API function calls and their arguments. The 16th CSI International Symposium on Artificial Intelligence and Signal Processing (AISP 2012), IEEE, 2012, pp. 563–568
32. C. Guarnieri, A. Tanasi, J. Bremer, M. Schloesser, The Cuckoo Sandbox, 2012

Chapter 4
Feature-Based Blood Vessel Structure Rapid Matching and Support Vector Machine-Based Sclera Recognition with Effective Sclera Segmentation

Chih-Peng Fan, Ting-Wing Gu, and Sheng-Yu He

4.1 Introduction

In recent years, the individual identification problems have become increasingly important for security applications. In the developed technologies for identity identification, the human biometric-based identity systems are safer and less prone to counterfeiting than the commonly used inductive card readers. Biometric information indicates metrics, which are related to characteristics of human, and the authentication technologies of biometrics have been widely used in computer science for identification and access control [1]. Many previous studies [2, 3] have shown that biometric features of humans are unique. In several human characteristics, the biometric features of eyes are also highly unique and protective by eyelids, and the eye features are not susceptible to external factors. Figure 4.1 reveals the diagram of eye structure and scleral area. Some previous sclera recognition designs [5–8, 12–14] were developed for identity identification, and Fig. 4.2 shows the general design flow of identity identification by using sclera blood veins. In [4], the sclera in eyes is the region, which is defined as the composition of the white and external opaque protective layer, and the sclera includes four layers of tissue: (1) the base layer, (2) the thin layer, (3) the sclera outer layer, and (4) the endothelium surrounding the iris. The visible blood veins are randomly distributed in the sclera region, and even the twins have different vein textures in the sclera region. In the sclera region, blood veins are visible and stable over time. Everyone's vein texture in sclera is randomly formed. As the age increases, collagen and elastin deteriorate, and the glycosaminoglycan loss causes scleral dehydration, and then lipids and calcium salts are also accumulated in sclera region. However, the vascular

C.-P. Fan (✉) · T.-W. Gu · S.-Y. He
Department of Electrical Engineering, National Chung Hsing University, Taiwan, Republic of China
e-mail: cpfan@nchu.edu.tw; g105064051@mail.nchu.edu.tw

© Springer Nature Switzerland AG 2019
G.R. Sinha (ed.), *Advances in Biometrics*,
https://doi.org/10.1007/978-3-030-30436-2_4

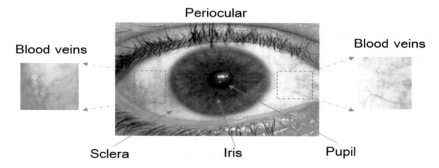

Fig. 4.1 The diagram of eye structure and scleral area

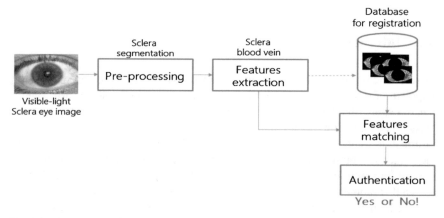

Fig. 4.2 The general design flow of identity identification system by sclera blood veins

textures in sclera do not deteriorate and change. For human beings, the genetic and developmental components determine the unique structure of their vascular patterns in the sclera. Besides, the texture of scleral blood vessel is also obtained by the noninvasive capture with visible wavelength illumination. Thus, the pattern of blood vessel in sclera is applied for identification by processing visible-light eye images. However, how do we perform the scleral recognition? Can we achieve high recognition accuracy by sclera veins? Based on visible-light eye images, a sclera recognition system can be developed by effective sclera blood vein extraction, and the matching efficiency of blood vein features will become accurate.

For sclera segmentation, the sclera region can be segmented straightforwardly by manual process, and the sclera region can be segmented quite accurately. However, the manual process usually requires artificial supervision and pays a large amount of time cost. For the real-time application issue, it is not feasible to cut the sclera region by manual operations, so the proposed design focuses on the development of automatic scleral separation. Before the process of sclera separation, the iris separation must be done usually. Daugman [9, 10] proposed the high-accurate iris segmentation algorithm, and the method was based on the high contrast characteristics between the iris and scleral boundaries, and then the two circles corresponding to the iris and sclera boundaries were estimated. By exhausting all possible circumferences and subtracting the accumulated values to get the largest difference, the iris region was found correctly. In [13, 14], the systems pre-located the center position of the pupil or the position of the iris before cutting the sclera region. Since the iris and the sclera regions in eyes have relatively fixed locations, finding these positions in advance will effectively improve the accuracy of extracting the sclera zone in eyes. In [15], the binarization process was used to select the possible region of the iris, that is, the region of interest (ROI). Then, the Canny filtering process was applied to find the edge of iris, and the pupil was located by Hough circular transform to estimate the position of pupil's center. In [16], the authors also proposed another methodology, which was divided into two operational phases. In the first phase, the process was based on the gradient method to generate a binary edge map. In the second phase, Hough transform concept was used for iris segmentation. Compared with Daugman's algorithm in [10], the main contribution in [16] was to realize the iris contour segmentation, which eliminated the interference objects in segmentation process, such as eyelashes and eyelids. In [17], the authors also proposed the useful algorithm, including pupil edge detection, pupil localization, reflection elimination, area filling, and iris boundary detection. In [17], the experimental results revealed that the accuracy of iris segmentation was up to 98%. By experiments, the method by Hough circular transform is less adaptable to eye images in various conditions, and the required value of threshold will be different for various shooting environments and iris radius. To select the appropriate threshold, it may not be suitable for uses in a fully automated identity identification system with the sclera. Besides, the method in [17] proposed the iris segmentation algorithm with higher accuracy than the Daugman's method. However, owing to the high complexity in [17], the method was not easy to be used for real-time applications. The suitable sclera-based identity identification system not only has the requirement of high recognition accuracy but also has less computing time for real-time applications. In this chapter, the developed iris segmentation methodology is optimized by using the Daugman's algorithm to quickly segment the iris.

In [18–20], the previous researchers used the scleral characteristics in different color spaces, and the eye images were converted into distinct color spaces, e.g., RGB, HSI, HSV, and YCbCr, to construct a scleral mask. Regardless of the applied color space conversions, these skills were based on the detection for non-skin area, and the technology separated the eye region by using skin color, and then it removed the iris region through the iris mask to obtain the scleral area successfully. In natural light environments, the eye images are easy to have a reflective effect. In the RGB color space, the reflective points in the sclera region have high pixel values, which may cause serious noises to affect the scleral cutting efficiency. In [21], the method of removing the reflective points was added as the pre-processing to cut the sclera region. The gray-scale scleral image was used to find the reflective area by calculating the histogram of the image, and then the bilinear interpolation method was used to fill the noise points. Due to the rapid development of high-performance graphic processors, the deep learning technologies have followed the trend, and the deep learning methods have outstanding performance in applications of object recognition and classification. Researchers in various fields have tried to introduce deep learning skills to increase the system performance. For identity identification by eyes, the method in [22] used the conditional random field (i.e., CRF) and convolutional neural network (i.e., CNN) technologies, and the recognition accuracy of the system was up to 83.2%. In [23], the authors applied the Segnet network to divide the contour of the iris. Since the textures of blood veins in the sclera are very subtle, to increase the identification accuracy, it is necessary to strengthen the vein textures. In [21], the authors used the Gabor filter to extract vein features, and the contrast-limited adaptive histogram technology was used to enhance the vein features. In [11, 24], by the gray-scale histogram in eye images, a clearer vascular model was obtained by the top-hat transform technology.

The others of this chapter are arranged in the following. In Sect. 4.2, the proposed methodologies are described for identity identification by sclera blood veins. Section 4.3 shows the experimental results and comparisons. Finally, a conclusion is stated at the end of the chapter.

4.2 Proposed Design Methodologies

The developed system uses the two-stage computations. The pre-processing process is used at the first stage, and the second stage is provided by two classification schemes, which are the K-d tree-based vascular feature matching identifier and the SVM-based classifier. Figure 4.3 depicts the computational flow of the proposed sclera identity identification system, and the applied two-stage computations for system design are depicted in the following.

4.2.1 Pre-processing Process

In the proposed computing flow, the first stage of the pre-processing process involves the segmentation of iris, the segmentation of sclera, and the enhancement of sclera blood vessel. At first, the improved Daugman algorithm is used in the proposed design for the iris fast positioning. Next, by using the segmented circle of iris, the system quickly segments the sclera region by means of the color information and the binarization process. Then the system enhances the sclera blood vessel features by the cascaded image processing, which includes the top-hat transform processing [11, 25], the contrast-limited adaptive histogram equalization (CLAHE) [26], and the process of Gabor filtering [27]. The details of pre-processing process are descried as follows.

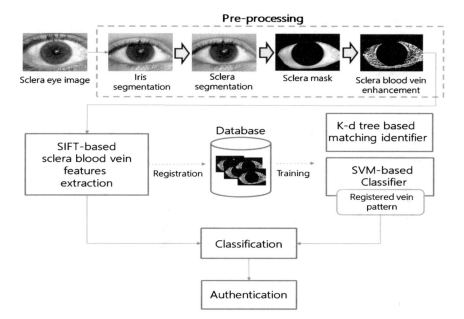

Fig. 4.3 The computational flow of the proposed sclera identity identification system

4.2.1.1 Iris Segmentation

To start the computational flow, the proposed system removes the reflective-light zone in the eye region. The reflective brightness on eyeballs leads to bright spots at local, and the color of the eyeball region is dark brown. Thus, the reflective light zone is revealed as white color. However, the brightness of bright speckle is similar with the brightness of scleral area, and then the reflective influence drops the iris segmentation accuracy. To achieve the iris location and segmentation, Daugman [10] proposed the effective algorithm, which is shown in Fig. 4.4. By exhaustively computing differences of accumulated value on all possible circumferences, the design [10] chooses the candidate circle of eye images, and the iris position will be estimated. In [13, 20], the system applies the modified Daugman's method, which has less computational complexities than the previous method in [10]. To reduce the amounts of calculations, the fast iris segmentation in [19] uses the three effective schemes as follows: (1) pre-select the centers of circles for calculating the eye area, (2) reduce the exhaustive computations in circumferences, and (3) reduce the image resolution by downsampling. In Fig. 4.5a, the pre-selected region of interest (ROI) of eyeball center is obtained by shrinking this entire image toward the center. The proposed system uses the fast computational schemes to reduce the number of candidate centers to be searched and also reduce the number of circumferential sampling points, where the schemes are shown in Fig. 4.5b and Fig. 4.5c, respectively. In the process of iris searching, to avoid the interference

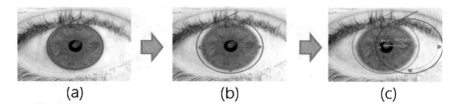

Fig. 4.4 The method proposed by Daugman [10]. (**a**) Calculate the gray-scale accumulated value on the circumference. (**b**) Subtract accumulated values of adjacent circumference. (**c**) Calculate the circumferential accumulated difference of all possible candidate centers

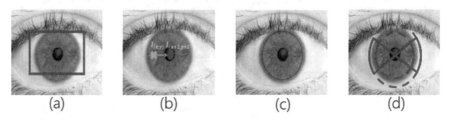

Fig. 4.5 The applied method in [20]. (**a**) Pre-selected region of interest (ROI). (**b**) Reduce the number of candidate centers to be searched. (**c**) Reduce the number of circumferential sampling points. (**d**) Select partial fan-shaped circumference for sampling

of upper and lower eyelids or eyelashes, only the fan-shaped pixels on both right and left sides of the circumference, as shown in Fig. 4.5d, are employed to compute the accumulated values for finding the iris center. By using the abovementioned schemes, the iris positioning becomes fast and precise.

By using the similar methodologies in [13, 20], the iris segmentation in [14, 21] includes the additional processing for pupil's location. When the iris segmentation is enabled, firstly the design [14] locates the center coordinate of pupil to reduce the computational complexity and to raise the segmentation accuracy simultaneously. In [21], the experiments show the contrast between the iris and pupil is very high, if the eye image is presented by the V-channel image with the HSV color space. Therefore, before the pupil location is processed, the eye image with the RGB color space is converted to the images with the HSV color space, and the eye image by the V-channel is taken as the input image. Besides, the operations for locating the center of pupil can be as low as possible. By referring to the method in [17], the searching range of region of ROI for pupil's location is reduced further, and the selected ROI region will contain the pupil, and then the process also reduces many extra calculations.

4.2.1.2 Sclera Segmentation

Figure 4.6 illustrates the computational flow of the proposed sclera segmentation method in [13]. At first, the gray-level eye image is processed by the smoothing

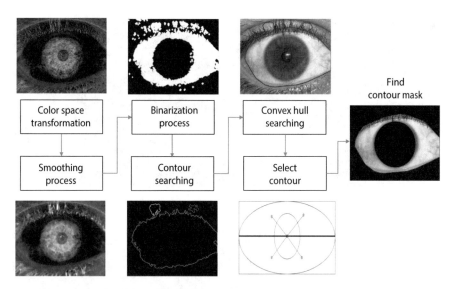

Fig. 4.6 The processing flow of the applied sclera segmentation

process. Next, the gray-level eye image is processed by the binarization process, and then the system estimates the contour of the sclera region. By the contour search, the convex hull of sclera region will be created. Based on the mask of contour, the region of sclera is recovered successfully. In the sclera segmentation process [14], by utilizing the low saturation property in the HSV color space, the sclera contour is achieved effectively, and the convex hull of sclera region is extracted properly.

4.2.1.3 Sclera Blood Vein Enhancement

Before the image enhancement is done, textures and features of blood veins on the sclera region are not obvious, and the eye images are also noisy. After the sclera region is processed by image enhancement, the textures of blood veins become more visible, and the sclera image restores the useful information of vein textures and features. In general, the sclera blood veins, presented by the green channel image, contain clear visible vein features. Therefore, the green channel image is utilized for image enhancement by the following cascaded processes: the top-hat transform [25], the contrast-limited adaptive histogram equalization (i.e., CLAHE) [26], and Gabor filtering processes [27]. Firstly, the top-hat transform is utilized to enhance blood vessel textures, and the CLAHE process is followed to raise the enhancement effect of the sclera vein textures. To reinforce the vein textures more, the proposed design enables the process of Gabor filtering. Figures 4.7, 4.8, and 4.9 illustrate the results after the top-hat transform, CLAHE, and Gabor filtering process, respectively [21].

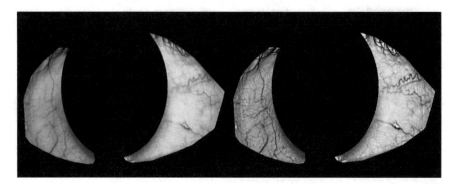

Fig. 4.7 The result by top-hat transform processing [21]

Fig. 4.8 The result by the
CLAHE processing [21]

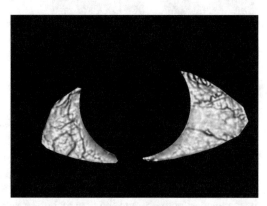

Fig. 4.9 The result by Gabor
filtering process [21]

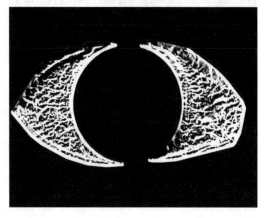

4.2.2   *Features Extraction*

In order to extract the blood vessel features in sclera, the local feature descriptors
used in [13, 14] provide the key directional information after the enhancement

of vein textures. The scale-invariant feature transform (i.e., SIFT) local feature descriptors are able to conquer the obstructions, which are generated by the variations of angles, environments, and distances when the images are shooting. Besides, the obstructions also involve angle, scale, bright, and displacement changes in shooting images. In [12], the local feature descriptors of sclera veins are extracted by the dense-SIFT algorithm, and the dense-SIFT technology estimates the local feature points effectively.

4.2.3   Features Training and Classification

4.2.3.1   By K-d Tree-Based Matching Identifier

By feature-based blood vessel structure rapid matching design [13, 20], the density samples are processed continuously. At the neighborhood in each feature point, the dense-SIFT-related information is computed, and then the system enables the structural vessel feature matching process. Besides, the system applies the random sample consensus (i.e., RANSAC) process to raise the efficiency of matching, and the corresponding relationship on the geometry is used to find the optimal matching couple points, and then the system effectively improves the accuracy of recognition. The detailed procedures by using the K-d tree-based matching identifier are described as follows.

When the calculations are processed for extracting the information of dense-SIFT features, supposing that the system uses the process of direct and exhaustive feature matching, all possible matching operations must compute the Euclidean distance, which is ranged between two vessel feature vectors. In tests, the immediate pairing skill usually decreases the relation of structural similarity corresponding to the feature sites between two sclera images. Thus, the proposed system uses the K-d tree algorithm [28] to reconstruct a binary search tree for the vein structure of scleral zone, and the scheme of features matching seeks the nearest neighbor of vein features. Figure 4.10 depicts the concept of the K-d tree processing. The applied K-d tree scheme speeds up the searching operations if the number of vein features is extensive. By searching the nearest neighbor features of scleral veins, the estimated matching pair cannot own the smallest distance. Thus, the used K-d tree operations construct the relative relationship, which is between the similar feature scheme of sclera veins, and the matching process is finished by a certain structural features of blood vessels.

By tests, the matching and probing time by the brute force algorithm is the same as the searching time by the K-d tree algorithm. However, the accuracy of recognition by the K-d tree methodology is better than that by the brute force method. Therefore, the proposed design exploits the K-d tree related nearest neighbor search technology for features matching, and the feature sets with more vein features will be created. Owing to the RANSAC process [29], the largest inner sets are acquired, and the outer sets are excluded, where the outer sets do not

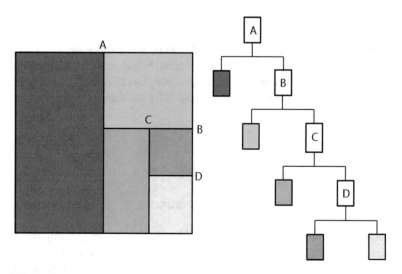

Fig. 4.10 The diagram for describing the K-d tree processing

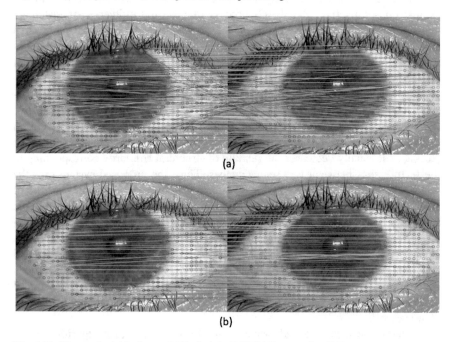

Fig. 4.11 The results (**a**) before and (**b**) after the RANSAC processing [20]

close the spatial correlation of vein features, and the features matching efficiency is improved. Figure 4.11a, b illustrates the results of before/after using the RANSAC processing, respectively.

4.2.3.2 By SVM Classifier

By the effective K-means method, the similar features are merged together by the proposed SVM-based design [14, 21] to set up a dictionary for depicting the features of interested group. Next, the images of sclera region consult the dictionary to obtain the histogram of group features, and the system feeds the group features to the support vector machine for training an identity classifier, and then the performance of sclera recognition is observed. The detailed procedures by the SVM classifier are described as follows.

By the SIFT-based algorithm [12], the system can extract the blood vein features effectively. By the extracted sclera vein features, the bag-of-features method is used for features combination to construct the K-means [30]-based features dictionary. Figure 4.12 shows the K-means-based bag-of-features methodology [21], which is used in the proposed design. The bag-of-features algorithm deposes the characteristics of all training data into a bag, and the system uses the K-means-based methodology to join similar vein features in the bag to compose a dictionary. The number of vocabularies in the dictionary is "K," which is called the word count, and the word count is set to 35 in the proposed design. Based on the K-means algorithm, the developed system collects the same blood vessel features together to build a characteristic wordbook for the features of interested veins. Then the images of sclera zone refer the dictionary of bag-of-features to obtain the main vessel features histogram, and the SVM scheme feeds the main vessel features for training a suitable classifier efficiently.

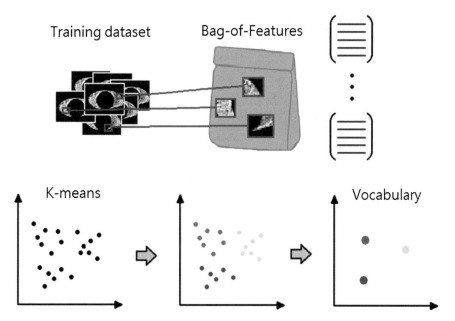

Fig. 4.12 The K-means-based bag-of-features methodology [21]

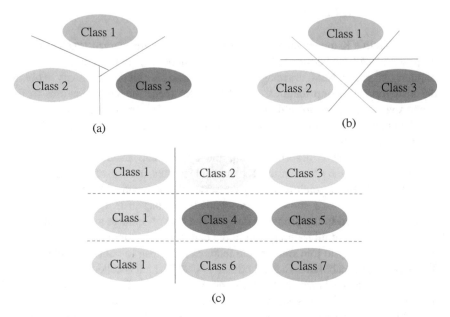

Fig. 4.13 (**a**) One-to-one SVM classifier. (**b**) One-to-all SVM classifier. (**c**) The used one-to-group SVM classifier

In general, the well-known SVM classifier [31] has two classification modes, where one is the one-to-one mode as shown in Fig. 4.13a and the other is the one-to-all mode as shown in Fig. 4.13b. In our design, when the SVM classifier operates in the one-to-one mode, the false positive rate cannot be decreased. On the other hand, if the SVM classifier operates in the one-to-many mode, the performance of recognition accuracy is not good enough since many negative samples are included in the training phase. Figure 4.13c depicts the used one-to-group SVM-based classification scheme. To train the proposed SVM classifier, the training procedure is conducted in groups, and each training group is separated into five categories. For training the proposed SVM classifier of Class 1, the Class 1 appears in each class, and each class has its one-to-group SVM classifier. In the testing procedure, the one-to-group classifier, which has the highest matching score, provides the output result for identity identification.

4.3   Experimental Results and Performance Comparisons

To verify the proposed design, the UBIRIS database [32] is used for the performance test. In the database, open source images are separated by two groups, where the total number of images is 1877, and the database provides high-quality (i.e., 800×600) and low-quality (i.e., 200×150) color images with JPEG format. The

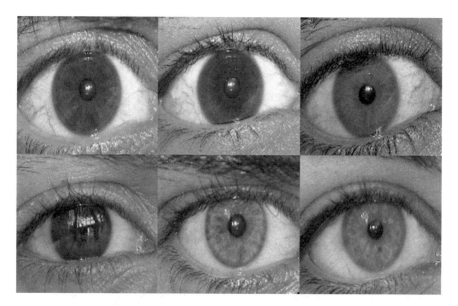

Fig. 4.14 Some selected high-quality images in UBIRIS database [32]

UBIRIS database was taken by 241 volunteers. Figure 4.14 illustrates several high-quality images in the UBIRIS database. The images at the first raw are selected from the first group, and those at the second raw are selected from the second group.

The most significant feature of the UBIRIS database is the addition of noise factors. For example, from Fig. 4.14, there are many natural light reflections around the region of eyes, and this effect achieves more realistic environmental conditions. In the first group, 1214 images are taken in a dimly lit room with some lighting equipment to minimize the noise of reflected light or other bright light. In addition, 663 images in the second group change the shooting condition, and the second group images are focused on the addition of natural lighting factors, which also cause the images to show the reflection, contrast, brightness, focus effects, and so on. Compared with the first group images, the images in the second group change the shooting angle and adjust the artificial shooting sources to increase variability. In Fig. 4.14, the high-quality images with more vascular lines and clearer images are suitable for the application of identity identification. However, the low-quality images with out-of-focus, less vascular information, eyelid interference, or closed eyes condition are not selected to test and verify the proposed design. Figure 4.15 shows some dropped images, which are not used for the experiments.

In the experimental environment, the personal computer has an Intel CPU with 3.40 GHz operational frequency and 8GB memories, and the software system model is executed in the 64-bit operation system. The C/C++ compiler with Visual Studio 2013 and OpenCV 2.4.13 [33] software library is used to develop the proposed software model. In addition, the applied UBIRISv1 sclera image database includes 241 persons with a total of 1214 sclera images, and each individual owns five

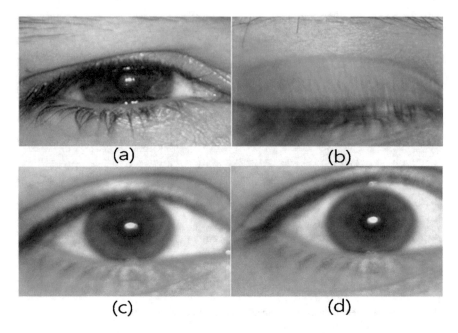

Fig. 4.15 The dropped images [32]: (**a**) Covered by eyelids and eyelashes. (**b**) Closed eyes. (**c**) Image blur caused by out-of-focus shooting. (**d**) Eyes are shot at the inappropriate position

corresponding eye images. The process of sclera segmentation will be affected by interferences of light, skin color, eyelids, and eyelashes. In the experiments, if the area of the extracted sclera mask covers more than 80% real sclera mask in an eye image, the system defines the estimated sclera mask is correctly segmented. Figure 4.16 illustrates the results of sclera segmentation in various cases by the proposed method in [14]. In Fig. 4.16, the results show that the sclera-based identity identification cannot be performed correctly in the case of eye diseases. In Table 4.1, by the developed pre-processing processes, the accuracies of sclera segmentation in [13, 14] are up to 95.85% and 98.35%, respectively.

Usually, the outputs of the biometric identification system are continuous values, and then the suitable threshold must be established to distinguish the four possible conditions. Table 4.2 describes the measurement matching matrix, which is also called the confusion matrix [34], where the four kinds of output possibilities are true negative (i.e., TN), true positive (i.e., TP), false negative (i.e., FN), and false positive (i.e., FP). In the testing phase, the sclera image samples in the database are divided into legal users and intruders. Then the false acceptance rates (i.e., FAR), false rejection rates (i.e., FRR), and equal error rate (i.e., EER) are used to evaluate and test the proposed design for performance comparisons. By the experiments, both of the FAR and the FRR values are less than 3%. Figure 4.17 illustrates the ROC curve in [21], and the EER value of our method in [21] approximates those of the previous designs in [6, 7].

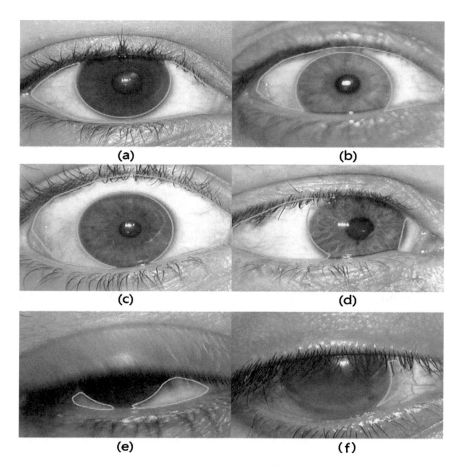

Fig. 4.16 Results of sclera segmentation by the proposed method in [14, 21]. (**a**) Normal. (**b**) Image defocus. (**c**) Eye enlargement. (**d**) Eyeball shifts to the corner of the eye. (**e**) Eyelid obscuration. (**f**) Eye disease

Table 4.1 The comparison of execution time and sclera segmentation accuracy in the three methods

	Execution time (seconds)		
Methods	Alkassar [7]	Method in Sect. 4.2.3.1	Method in Sect. 4.2.3.2
Pupil location	N/A	N/A	0.004
Iris segmentation	1.97	0.082	0.062
Sclera segmentation	0.247	0.037	0.074
Total time	2.217	0.119	0.140
Segmentation accuracy			
Sclera segmentation accuracy	98.65%	95.85%	98.35%

Table 4.2 Measurement matching matrix

		True class	
		Positive	Negative
Predicted class	Positive	True positive (TP)	False positive (FP)
	Negative	False negative (FN)	True negative (TN)

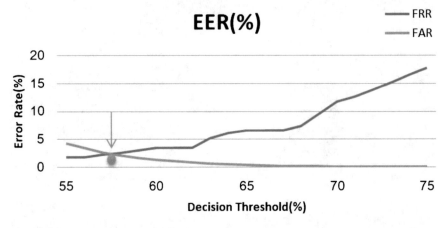

Fig. 4.17 The ROC curve

Table 4.3 The recognition accuracy of the proposed SVM-based design

Number of testing categories	Number of images for training/number of images for testing	Accuracy of identity recognition
31	124/31	98.33%
41	164/41	~ 100%
61	244/61	~ 100%
81	324/81	~ 100%

In [14], the developed one-to-group SVM classifier identifies the tested eye image which category is closest to the database, not whether it is a legitimate user or an illegal intruder, and then the identification output is a binary result. In [14], for performance measurements, the testing results of true positives and false negatives are classified into the correct predictions. On the contrary, the testing results of true negatives and false positives are classified as the prediction errors. Table 4.3 shows the recognition accuracy of the proposed SVM-based design in [14]. When the number of categories increases to more than 41, the recognition accuracy of identity identification will approach 100%.

4.4 Conclusions

In proposed design, the system integrates the process of adaptive histogram equalization, the operation of Gabor process, and the technology of fast features matching to enhance the efficiency of recognition. For identity identification, the system uses the developed dense-SIFT-related rapid sclera vessel matching skill. By tests with the UBIRISv1 dataset, the accuracy of recognition of the features matching-based method can be up to 96%. Besides, the values of the FAR and the FRR are smaller than 3%, and the value of EER approximates that of the previous design in [7].

By using SIFT features of sclera blood vessels, the sclera recognition system with effective machine learning technology is also developed to the application of identity identification. After the sclera segmentation and features enhancements by the K-means-based histogram of group vein features, the developed one-to-group SVM scheme is used for identity recognition after training. By experiments with the UBIRISv1 dataset, the recognition accuracy of the proposed SVM classifier can be up to near 100%.

References

1. https://en.wikipedia.org/wiki/Biometrics
2. P.M. Corcoran, Biometrics and consumer electronics: A brave new world or the road to dystopia? [Soapbox]. IEEE Consum. Electron. Mag. **2**(2), 22–33 (2013)
3. J.A. Unar, W.C. Seng, A. Abbasi, A review of biometric technology along with trends and prospects. Pattern Recogn. **47**, 2673–2688 (2014)
4. https://en.wikipedia.org/wiki/Sclera
5. Z. Zhou, E.Y. Du, N.L. Thomas, E.J. Delp, A new human identification method: Sclera recognition. IEEE Trans. Syst. Man Cybern. Syst. Hum. **42**(3), 571–583 (2011)
6. Y. Lin, E.Y. Du, Z. Zhou, N.L. Thomas, An efficient parallel approach for sclera vein recognition. IEEE Trans. Inf. Forensics Secur. **9**(2), 147–157 (2014)
7. S. Alkassar, W.L. Woo, S.S. Dlay, J.A. Chambers, Robust sclera recognition system with novel sclera segmentation and validation techniques. IEEE Trans. Syst. **47**, 474–486 (2017)
8. S. Alkassar et al., Sclera recognition: On the quality measure and segmentation of degraded images captured under relaxed imaging conditions. IET Biom. **6**(4), 266–275 (2016)
9. J.G. Daugman, How iris recognition works. IEEE Trans. Circuits Syst. Video Technol. **14**(1), 21–30 (2004)
10. J.G. Daugman, High confidence visual recognition of persons by a test of statistical independence. IEEE Trans. Pattern Anal. Mach. Intell. **15**, 1148–1161 (1993)
11. X. Bai, F. Zhou, B. Xue, Image enhancement using multi scale image features extracted by top-hat transform. Opt. Laser Technol. **44**(2), 328–336 (2012)
12. D.U. Pal, M.A.F. Ballester, M. Blumenstein, Sclera recognition using Dense-SIFT. 13th International Conference on Intelligent Systems Design and Applications, 2013, pp. 74–79

13. T.W. Gu, C.P. Fan, Sclera recognition by density sampling features based vascular structure rapid matching for identity identification. IEEE 7th Global Conference on Consumer Electronics, Nara, 2018
14. S.Y. He, C.P. Fan, SIFT features and SVM learning based sclera recognition method with efficient sclera segmentation for identity identification. The 1st IEEE International Conference on Artificial Intelligence Circuits and Systems, Taiwan, 2019
15. L.V. Romaguera, et al., Pupil segmentation approach on low resolution images. VI Latin American Congress on Biomedical Engineering CLAIB 2014, Paraná, Argentina, Springer, Cham, 2015
16. R.P. Wildes, Iris recognition: An emerging biometric technology. Proc. IEEE **85**(9), 1348–1363 (1997)
17. A.T. Hashim, D.A. Noori, An approach of noisy color iris segmentation based on hybrid image processing techniques. 2016 International Conference on Cyberworlds, 2016
18. S. Alkassar, et al., A novel method for sclera recognition with images captured on-the-move and at-a-distance. IEEE 4th International Workshop Biometrics and Forensics, 2016
19. S. Alkassar, et al., Enhanced segmentation and complex-sclera features for human recognition with unconstrained visible-wavelength imaging. IEEE International Conference Biometrics, 2016
20. T.W. Gu, Sclera recognition by density sampling features based vascular structure rapid matching for identity identification, Master thesis, Department of Electrical Engineering, National Chung Hsing University, 2017
21. S.Y. He, SIFT features and machine learning based sclera recognition method with efficient sclera segmentation for identity identification, Master thesis, Department of Electrical Engineering, National Chung Hsing University, 2018
22. M. Russel, B. McCane, S. Mills, Conditional random fields incorporate convolutional neural networks for human eye sclera semantic segmentation. IEEE International Joint Conference on Biometrics, 2017
23. N. Sinha, A. Joshi, A. Gangwar, A. Bhise, Iris segmentation using deep neural networks. IEEE 2nd International Conference Convergence in Technology, 2017. https://doi.org/10.1109/I2CT.2017.8226190
24. D. Gou, T. Ma, Y. Wei, A novel retinal vessel extraction method based on dynamic scales allocation. 2nd International Conference on Image, Vision and Computing, 2017
25. https://en.wikipedia.org/wiki/Top-hat_transform
26. https://en.wikipedia.org/wiki/Adaptive_histogram_equalization
27. https://en.wikipedia.org/wiki/Gabor_filter
28. https://en.wikipedia.org/wiki/K-d_tree
29. https://en.wikipedia.org/wiki/Random_sample_consensus
30. D. Arthur, S. Vassilvitskii, k-means++: The advantages of careful seeding. Proceedings of the eighteenth annual ACM-SIAM symposium on Discrete algorithms, Society for Industrial and Applied Mathematics, Philadelphia, PA, USA, 2007
31. "Support Vector Machines for classification," http://efavdb.com/svm-classification/
32. H. Proenca, L.A. Alexandre, UBIRIS: A noisy iris image database, in *Image Analysis and Processing-ICIAP*, (Springer, Berlin, 2005), pp. 970–977
33. http://opencv.org/
34. https://en.wikipedia.org/wiki/Confusion_matrix

Chapter 5
Different Parameter Analysis of Class-1 Generation-2 (C1G2) RFID System Using GNU Radio

Parvathy Arulmozhi, P. Varshini, Pethuru Raj,
John Bosco Balaguru Rayappan, and Rengarajan Amirtharajan

5.1 Introduction

The fast growth in RFID communication has demonstrated that remote communication is feasible for information service. By using a specific standard, conventional remote devices are structured to transfer an individual communication service [1]. The cost of the hardware devices is expensive and not user-friendly. In an RFID system, it is more tedious to work with hardware implementation [1–5]. The readers in the RFID system are suitable to transmit and receive radio signals only for low (LF) and high frequencies (HF). Hence it is not possible to transmit RFID signals in the conventional hardware system above that range. Thus, the implementation of the traditional system of hardware is more arduous and not flexible for the users. A viable solution to make a communication system more flexible can be accomplished through software implementation. It provides a user-friendly environment. The technology developed for the software implementation is a software-defined radio concept known as SDR. SDR is one of the most recent techniques developed for modern wireless-based communication systems. Thus, the same hardware setup can be used to create various radios for various transmission and reception standards. This unique platform performs various techniques like modulation and demodulation of the signals at different frequencies, narrowband/wideband operation, etc., thus providing an easy adaption. Therefore, it diminishes the usage of hardware peripherals in the RFID system [1–5].

P. Arulmozhi · P. Varshini · J. B. B. Rayappan · R. Amirtharajan (✉)
School of Electrical & Electronics Engineering, SASTRA Deemed University, Thanjavur, Tamilnadu, India
e-mail: amir@ece.sastra.edu

P. Raj
Reliance JioInfocomm. Ltd (RJIL), Bangalore, India

© Springer Nature Switzerland AG 2019
G.R. Sinha (ed.), *Advances in Biometrics*,
https://doi.org/10.1007/978-3-030-30436-2_5

The software setup is implemented using GNU radio companion (GRC) platform. Eric Blossom created this GNU radio software under GNU general public license. The software is an open-source, and it is accessible in various OS platforms (Windows, Linux, etc.). It is a graphical user interface, analogous to the simulation platform. GNU radio is the platform mainly used to perform signal processing operations required for the hardware setup. The digital communication system can be implemented using this software. All the blocks in the GNU radio software are written in C++ and Python language.

There are various predefined elements, like modulators, demodulators, filters, encoders, etc., that are available in the software. These are known as blocks. Users can also embed their blocks by writing the respective Python code for their blocks to implement new functions [6–15]. More than 150 signal processing blocks are available in GNU radio software. Hence it is much more convenient to transmit RFID signals at ultrahigh frequency by using this GRC platform. New signal processing blocks can be easily embedded in this platform. Since it is a software platform, it handles only the digital input data. There is a beneficial tool in this GNU radio, which is called Simplified Wrapper and Interface Generator (Swig), required to change the classes used in C++ into the classes utilised in Python. Hence it is more beneficial to perform the signal processing for various hardware setup devices like SDR, USRP, etc.

In this chapter, we are presenting various modulation and demodulation techniques on the RFID signals for the tag sequence generation, utilised for the communication between a reader and a tag in the RFID system. Therefore, the performance of each modulation technique at various RFID frequencies can be analysed with their corresponding constellations, power and bit error rate [2].

5.2 RFID EPC C1G2 Protocol

5.2.1 Representation of RFID EPC Protocol in GNU Radio

Electronic Product Code or EPC is the name affiliated with ISO/IEC 18000-6C RFID standards [1]. This standard can be used only in the ultrahigh-frequency range (UHF). But this operates in the range 860–960 MHz. The carrier signal is generated and sent to the tag by the reader associated with the object to extract the information present in it.

Every product has a unique product code for easy identification. Hence the tag responds the reader with the product code associated with it. The reader sends the information to the database, which allows storage and interpretation of tag data in many other applications.

The Electronic Product Code consists of 64/96/128 bits long. The header, EPC manager, object class and serial number are 8 bits, 28 bits, 24 bits and 36 bits,

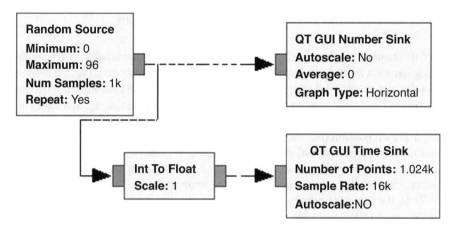

Fig. 5.1   Representation of 96-bit tag sequence using GNU radio

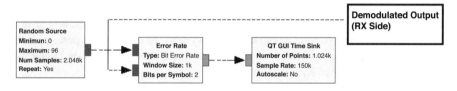

Fig. 5.2   BER representation in GNU radio

respectively. This is the representation of a specific product code, which is 96 bits long and represented as in Fig. 5.1.

This protocol has been introduced to send the information at a higher range of frequency band. During the transmission of the radio waves, the Electronic Product Code uses FHSS (frequency-hopping spread spectrum). It is a method that quickly switches the carrier signal among different frequency channels to obtain a possible read from the tag [3].

5.2.2   Representation of BER in GNU for C1G2 Protocol

Bit error rate can be represented using the error rate block in GNU radio. The error rate block determines the number of errors obtained in the system. Hence one of the inputs can be taken as the information source, and the other input is the demodulated output as depicted in Fig. 5.2.

5.3 Introduction to RFID Authentication Factor

In this chapter, we described various modulation and demodulation schemes for the generation of 96-bit tag sequence, which has to be implemented with a defined security algorithm to enhance RFID security [4]. As discussed before, we modulate our product code information associated in the tag with the carrier signal transmitted by the reader over the channel. When we transfer the data over the channel, there are a lot of possibilities for various attackers to extract our information while transmitting. Hence anyone can perform attacks on our original information. So, it is required to send in a secured manner, and the RFID tag and the reader should authenticate mutually with each other before sending data.

Thus, the necessary communication between the reader and tag should be verified by the user to eliminate malicious attacks in our RFID system and enhance privacy. If there are multiple tags in the reader range of RFID system, tag collision may occur. To avoid this, we can consider every tag sequence as a unique RFID authentication code and transmit over the channel. Since these authentication codes are different for various tags, it differentiates RFID tags by having dissimilar tag sequences which can prevent a collision in the system. To enhance the performance of the RFID system, it is required to implement various protocols and security algorithms for a secured RFID system.

In today's web and cloud applications, one of the most significant forces driven would be the online service. All the online services utilise the authentication for security and privacy of the user's concern. Whenever we request the services from any security system, authentication is required to verify the user's credentials to improve the security and privacy of the system design. Authentication can be either a single-factor or a multi-factor authentication.

Generally, single-factor authentication needs only one factor to log in to the system. For the RFID tags, the tag sequence itself can be considered as an authentication factor to verify if the tag sequence is a unique sequence number. To enhance the security algorithm, we can also implement the multi-factor algorithm. In the multifactor algorithm, we require more than one factor to log in to the system and access the information. These algorithms are implemented mainly to avoid system breaches. Thus the information is encrypted, and it is possible to gain access if and only if we know our key to access the secured data in a system.

The various applications of steganography can be considered as a replacement to the biometric systems. Each RFID tag sequence can act as a substitute for the existing biometric system. Thus, this can enhance the security in the system, and various attacks like brute force attack, dictionary attack and password guessing and tag number guessing can be avoided.

5.3.1 Single-Factor Authentication

Authentication is a method of verifying the credentials of the users. It can be verified using a single-factor authentication algorithm [4]. As mentioned earlier, only one factor is required for accessing the secured information. In a system, the username and the password altogether can be considered as a single-factor authentication in the system. Once the users provide their correct username and password, the system verifies and grants access if it matches the given password and the username. Else, the information cannot be accessed. Thus accessing the required information is strictly secured and restricted to avoid malicious attacks.

In single-factor authentication, the system depends only on one factor to provide the secured information to the client. The client gains access to the secured information if they contribute their credentials correctly. Hence, it can be considered as one-factor authentication. Authentication depending on password is a weak method which can be incorporated in the system since most of the attackers try guessing the password in their initial trials. But clients use passwords since it can be easily remembered. Also, they don't change their passwords frequently, which causes attackers to find out the password and corrupt the data easily. Hence to avoid all those, it is necessary to change their password regularly. To enhance security, we can go for multi-factor authentication.

5.3.2 Multi-factor Authentication

Biometric is one of the examples of multi-factor authentication to access our secured information. The various other applications like ATM, voice recognition and face recognition can be considered as multi-factor authentication for our system design. In ATM, the physical card is one of the factors for the identification, and the other factor is the PIN code we enter to access the secured information. Hence the user's identification (ID) is unique since the account number and the PIN code are unique for every user. Thus the same algorithm is implemented in various other applications to enhance the security of user's information.

In biometric, more than one factor is required to gain access to the secured information. In a system, the username and the password can be considered as one factor. Various other factors can be considered for enhanced privacy in the system. The fingerprint impression is an added factor to verify the user and to provide the information only if the fingerprint matches with the registered fingerprint. Also, the verification code word can be considered as one factor which will get received from the client. The PIN code will be asked as an added factor to avoid user's identification from being attacked or hacked from malicious users. The RFID tag sequence or the RFID token number itself can act as an authentication factor. These added factors would provide more security to our information. The actual biometric information can be a fingerprint or iris scan or voice recognition. It verifies with the registered data and provides access. Biometric systems convert the scanned data to a string or a mathematical expression and confirm it with the registered data. The biometric device examines the characteristics and extracts their information, converts them into a string of data and compares it with the registered expressions. In an RFID system, the tag sequence is being transmitted between the reader and the tag. The reader uses the RFID tag itself as an authentication factor to extract the data and store the information in the database. The tag sequence should be unique.

In GNU radio, the random source block was implemented to generate a random sequence number for various tags. Also, various encryption algorithms can be implemented to transmit information in a secured manner. The encryption block can be introduced after modulating the information where the key should be given in a transmitter section. While decrypting the data after demodulation, the same key should be provided to extract the information from the transmitter. Hence the system is more secure by providing the key to encrypt and decrypt the information in an RFID system.

5.3.3 RFID Factor Authentication Application

RFID has been widely used in various technology applications as it is less expensive and small in physical size, which can be placed anywhere required [2, 5]. RFID is an electronic method of data transfer assisted through radio frequency waves. It can be utilised in various applications but not limited to automated teller machine (ATM), security system for industries, educational institutions, medical data protection, security for smart logistics, etc. [16–26]. It has been predominantly used for authentication and object identification through RFID intelligent proofing.

The main motive of RFID technology is to transfer the information from the tag and retrieve the information stored in the tag using the reader and store it in the database. To store this information securely, we adopt a single-factor and multi-factor authentication, as discussed earlier. This authentication factor application verifies the code used in the transmitter to decrypt it in the receiver side section. Hence this RFID factor authentication helps us to store the encrypted key for every single tag. Thus every tag can be identified uniquely as it varies with a unique key for various tokens. Once the RFID tag gets scanned by the user, the system waits for the user to enter the code word and sends it to the server after ciphering the code word. This is the encryption method adopted to enhance the security in the communication between the tag and the reader. The server checks with the decryption algorithm and converts the cipher to the original code word and verifies the information received. Thus, every token stores the key information to tighten the security in RFID communication.

The user has to request a new user account to register their biometric information. While registering, each user will have a unique tag in which the information can be stored. They will receive a unique tag sequence while registering the username and the user password. Whenever we want to access the data, the user should provide their credentials and the unique key correctly analogous to the registered information to gain access to their data. Thus, it improves more security for the transmission of our information between the tag and the reader. In case if the attacker stole the user's credentials, he cannot be able to access the data since the attacker is unaware of the unique code generated in the user's system. This is known as the binding process, which is introduced mainly to avoid attacks onto the system.

5.3.4   Biometric Hash Functions

Hash functions are functions which are used to map the arbitrary input information to a fixed size value of integers [5]. The values returned by the hash function are known as hash codes, hash values or hashes which can uniquely determine secret information. It works like a hash table that provides a data lookup table. This hash table is mainly used to find and eliminate the duplicate records given as an input. These functions are similar to the biometric fingerprint, checksum and lossy compression as it verifies our input information with the hashes available in the lookup table. If the key value of the input data is present already in the lookup table, then it provokes the collision of two similar records. Hence this input information can be eliminated. If the input value is unknown, then it can be assigned to an integer value and can deliberately get recovered.

A cache is also used which is analogous to the hash table. When a collision occurs, it discards the input key if the hash value already exists. Hence if many tags have the same sequence number, then it results in a collision, and the cache will eliminate all the duplicate key by storing the first key as a hash value. Hence collision can be avoided, and all the tag sequences have a unique key which reduces the possibility of attacking our secured system from malicious attackers. Thus various applications employ the hash function to enhance the data integrity and for data corruption detection and authentication purpose to improve the security in the designed RFID system.

5.4   Digital Modulation Scheme for RFID System

In the reader-tag communication, the carrier signal will be generated by the reader and sent to the tag in the RFID system. Once the tag receives the carrier signal from the reader, it modulates the information signal with the received carrier signal using a suitable modulation scheme. We use modulation techniques to send the information to a longer distance in the range of UHF frequency. They provide security and large capacity to carry our information since it's a high-frequency range. Hence, the best technique can be analysed by determining the signal to noise ratio using various digital modulation schemes like ASK, FSK, BPSK, QPSK and QAM.

5.4.1 Binary Amplitude Shift Keying (ASK)

This is a modulation technique used to send the information signal by varying the amplitude of the carrier signal but not the phase and frequency of the carrier signal. In GNU radio, we implemented ASK by generating a signal source for the carrier signal generated by the reader at 13.5 MHz/930 MHz frequency (waveform-sine) and another signal source for generating the message signal at the 1KHZ frequency (waveform-square). It sends the bit '1' with the presence of the carrier, and it sends '0' with the absence of the carrier by adding 'Add Const' block.

Hence both signals are modulated and displayed in 'QT GUI time sink' block to represent the ASK waveform in the time domain. The noise can be introduced when it is passed over the AWGN channel. To recover our original information, we neglect the noise by introducing an FIR filter to produce the finite number of samples. Here, we used a low-pass filter with a Hamming window. The clock recovery block can be used to recover the timing information of our transmitted signal. Therefore, the power can be measured from the FFT spectrum to obtain the SNR ratio. The basic advantage of this technique is its simple implementation and high bandwidth efficiency, but the bit error rate is high and prone to more noise.

5.4.2 Binary Frequency Shift Keying (BFSK)

In this technique, the information signal is sent by varying the frequency of the carrier signal generated by the reader. In GNU radio, we implemented binary FSK to transmit the signal at various frequencies. Therefore, two signal sources are generated at different carrier frequencies in the UHF range such as 13.5 MHz and 930 MHz. The information signal is generated in another signal source at a frequency of 1 kHZ. Hence, we multiply the information signal with two carrier signals, and the resultant FSK signal is generated by adding it together. Under the AWGN channel model, the noise is introduced along with the modulated signal. Hence the appropriate filter and clock recovery can be introduced to get back the original signal. The power spectrum can be analysed using FFT sink. The basic demerit of this technique is that it has lesser efficiency and power compared to other modulation schemes [6].

5.4.3 Phase Shift Key (PSK)

This technique is used to send the information signal by varying the phase of the carrier signal and keeping the phase and the frequency of carrier signal constant. There are various types in phase shift keying. It assigns a group of bits uniquely infinite number of phases. Some of them are explained in Sects. 5.4.3.1, 5.4.3.2, and 5.4.3.3.

5.4.3.1 Binary Phase Shift Key (BPSK)

In this technique, it encodes the information in two phases, such as 0° and 180°, whereas the amplitude and frequency should be kept constant. This technique is vigorous since it produces high distortion. Hence it is not preferable for the applications which require a high data rate. A random source block in GNU radio is used to generate the random number of samples, and the required information tag sequence can be generated.

The PSK modulation block can be utilised to modulate the phase of the carrier signal to the constellation points and the samples that have to be generated per second mentioned in that block. The constellation point is given as '2' for binary phase shift keying. Hence the modulating signal can be multiplied with the carrier signal to generate the modulated signal, and the power can be measured in frequency sink. The PSK demodulation block is introduced to demodulate the signal with the same constellation points [7, 8]. We used the throttle to limit the sample rate to avoid CPU blockage.

Also, a bandpass filter is used at the receiving side to filter the unwanted signal [7]. The respective power and the constellation can be analysed using frequency and the constellation sink. Hence the information is encoded in two phases at various sample rate.

5.4.3.2 Quadrature Phase Shift Key (QPSK)

This technique is used to encode two bits per symbol. Hence it transmits double the amount of information in the available bandwidth. Therefore the data rate of QPSK is higher than that of BPSK. The term quadrature implies that the phase of a signal is orthogonal to the other one. Hence the phase shift of the carrier signal can be represented at four different phases such as 90°, 180°, 270° and 360° for the corresponding four symbols. The error rate block is used mainly to identify the bit error rate of our information. In this, the reference signal can be used as our input signal, and it compares it with the demodulated output to find out the number of errors occurred while receiving the information after demodulation. The output can be observed in the time domain using time sink. The output on the constellation diagram represents four points. Hence it's quadrature phase shift keying. Here the bit error rate is comparatively lesser than that of the modulation as mentioned in earlier schemes [9].

The constellation point in a PSK modulator should be given as 8, 16 and 64 to obtain the corresponding 8PSK, 16PSK and 64 PSK modulation schemes. The similar constellation plots are observed using constellation sink in GNU radio.

5.4.3.3 Quadrature Amplitude Modulation (QAM)

QAM is a modulation technique that is both analogue and digital. It uses two carrier signal waves with the same frequency. If one signal is represented as a sine wave, the other could be the opposite of that corresponding waveform (i.e. cosine). Hence, the phase of those carrier signals varies by 90 °. It transmits the input signal by altering the amplitude of the respective carrier signals by using either amplitude shift keying (ASK) or amplitude modulation (AM). The power obtained using QAM is higher when compared to the modulation, as mentioned above, schemes, which can be interpreted in the corresponding frequency and constellation sink plots. By varying the constellation point to 4, 16, 32 and 64, it is possible to obtain 4-QAM, 16-QAM, 32-QAM, etc., and the output can be realised using constellation sink in GNU radio software. Bit error rate (BER) can be further calculated using an error rate block, which can be realised in the time domain.

5.4.3.4 Analysis of Digital Modulation Schemes over AWGN Channel

This technique is mainly adopted to analyse the response of the modulated signal over the additive white Gaussian noise (AWGN) channel model. Hence we introduced a channel model in various digital modulation schemes.

5.4.4 PSK over AWGN Channel

The input sequence is generated using a random source block, and it is given to the PSK modulator block whose constellation point should be mentioned as 2 since it's a BPSK modulation. The resultant modulated output is then given to the throttle which limits the throughput of the signal and passes onto the channel model.

This model allows the user to set the voltage of an AWGN noise source (), a (normalised) frequency offset (), a sample timing offset () and a seed () to randomise or make reproducible AWGN noise source. The noise voltage can be varied according to the range mentioned in the QT GUI range. Once the noise has been introduced into the channel, it is necessary to extract our original information by the demodulation process. So, PSK demodulator with the same specification as mentioned in the transmitter section should be incorporated.

The rational resample block is to interpolate or decimate the incoming signal to increase or reduce the sample rate. The resultant power is comparatively higher than the power obtained from the modulation scheme with no channel model. Hence the spectral efficiency is improved, and power can be observed using the FFT sink, and the constellation plot is obtained using constellation sink. The merit of this technique is that it gives better SNR ratio when compared to ASK or FSK modulation scheme.

5.4.4.1 QPSK over AWGN Channel

In GNU radio, we transmit the information signal whose constellation point is 4 over AWGN channel model [1]. Hence it is QPSK modulated. The resultant demodulated output can be obtained by connecting it to the PSK demodulator block. It is filtered using a low-pass filter and analyses the spectrum using frequency sink. While comparing the power level obtained in this technique with the output obtained in the absence of channel model, the power obtained in this scheme is quite better than that of the other model. Hence spectrum efficiency is improved. SNR can be increased by decimating it with proper filtering. Therefore the bit error rate gets reduced at a higher rate [10].

5.4.4.2 QAM over AWGN Channel

In QAM, the same technique can be adopted, and the power level obtained over the channel model has a better performance than that of the power obtained in its absence. The BER of this modulation scheme is obtained, which comparatively have more BER for fewer samples/sec. As we increase the samples per second, the BER is converging to a smaller value. The more the samples per second, the lesser the bit error rate. Hence the signal to noise ratio is better when samples per second are more for our system design.

5.4.4.3 ASK over AWGN Channel

In GNU radio, ASK modulation is implemented by generating a signal source for the carrier signal generated by the reader at 13.5 MHz/930 MHz frequency (waveform-sine) and another signal source for generating the message signal at a 1 kHZ frequency (waveform-square) as shown in Fig. 5.3. Both are modulated and transmitted over an AWGN channel where the noise can be introduced. At the receiver, it can be demodulated by recovering the input clock pulses. The results can be displayed in WX GUI time and frequency sink to display the obtained waveform at RFID frequency.

5.4.4.4 FSK over AWGN Channel

In GNU radio, we implemented binary FSK to transmit the signal at various frequencies, as shown in Fig. 5.4. Therefore, two signal sources are generated at different carrier frequencies in the UHF range such as 13.5 MHz and 930 MHz and modulated with the message signal. The modulated signal is then passed onto the AWGN channel, and the resultant FSK waveform is retrieved by demodulation.

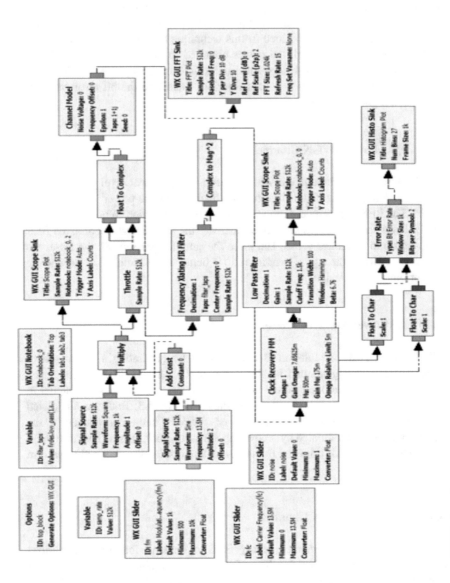

Fig. 5.3 ASK over AWGN channel

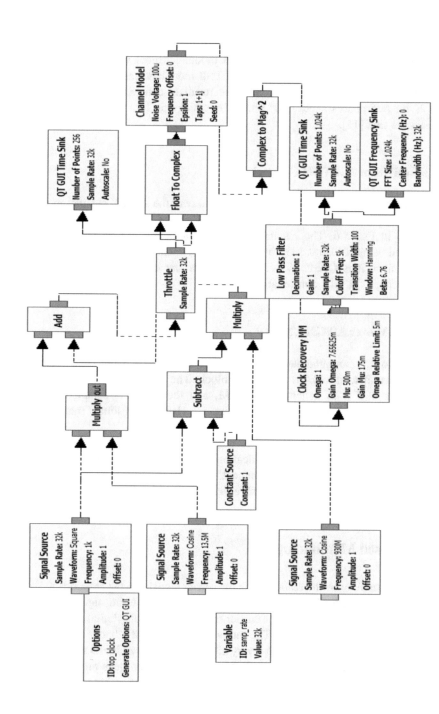

Fig. 5.4 FSK over AWGN channel

5.4.4.5 BPSK over AWGN Channel

The input sequence is generated using a random source block and encoded using differential encoder which is given to the PSK modulator block whose constellation point should be mentioned as '2' since it's a BPSK modulation as shown in Fig. 5.5. Once the demodulation process is done, the resultant constellation should have symbols on two phases such as 0° and 180° which can be seen using QT/WX constellation sink [11].

5.4.4.6 QPSK over AWGN Channel

In GNU radio, we transmit the information signal over the AWGN channel model, and the constellation point is mentioned as '4' since it is QPSK modulated as depicted in Fig. 5.6. The resultant demodulated output can be obtained and synchronised using a Costas loop block. The order mentioned in the Costas loop block is 4 since it is QPSK modulated. The resultant constellation plot can be displayed using a QT GUI constellation sink.

5.4.4.7 QAM over AWGN Channel

In GNU radio, we can generate a signal of 1 kHz frequency using signal source block and pass it to the QAM modulator block. The constellation point can be mentioned as 4 if it is generating 4-QAM, and the signal is sampled at the corresponding sample rate mentioned as in Fig. 5.7. Hence the resultant transmitted signal can be analysed in time sink.

At the receiver side, the modulated output can be demodulated by using QAM demodulator block with the same specification mentioned in the transmitter section. When it is passed over the AWGN channel, the BER is reduced as shown, thus providing an improvement in the spectral efficiency.

5.5 Proposed Methodology

In an RFID system, once the reader sends the carrier signal, the tag responds to the reader by transmitting the EPC associated with it as a backscattering signal [12, 13]. This signal can be modulated with the carrier signal using various digital modulation techniques, as mentioned Fig. 5.8. The motivation is to adopt a better modulation scheme to transfer the information at a very high frequency in an RFID system with less prone to noise. Thus, QPSK and QAM modulation techniques provide better SNR ratio. Hence the bit error rate and power can be analysed for these modulation schemes.

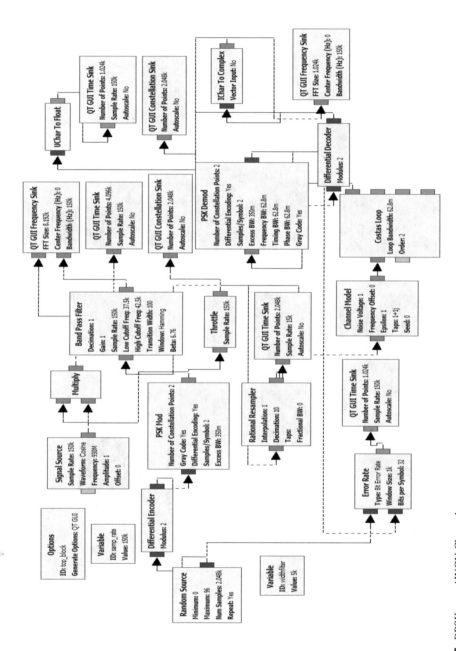

Fig. 5.5 BPSK over AWGN Channel

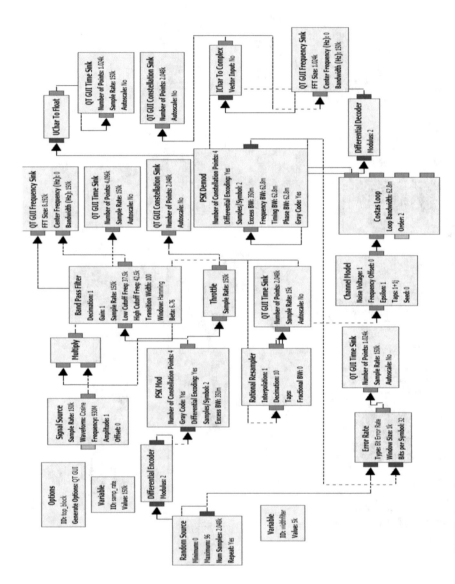

Fig. 5.6 QPSK over AWGN channel

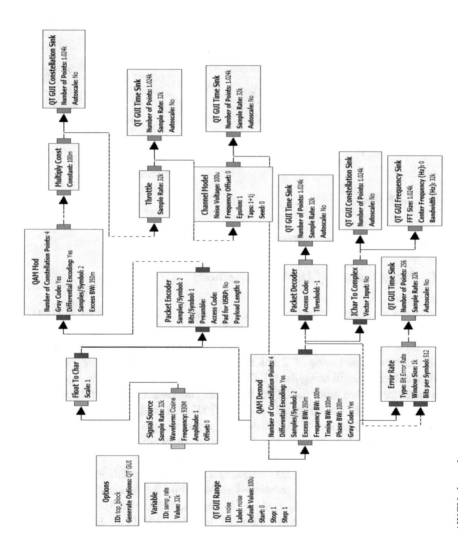

Fig. 5.7 4-QAM over AWGN channel

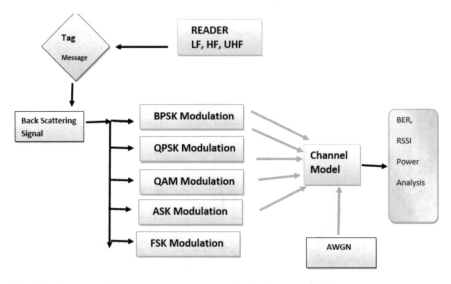

Fig. 5.8 System design

5.6   Results and Discussion

5.6.1   Performance of Detection Methods

The resultant output of various modulation schemes can be analysed using GNU radio software [14]. The output waveform and the power spectrum of ASK over the channel model can be shown using time sink as shown in Fig. 5.9, and the waveform of FSK can be obtained using FFT sink as shown in Fig. 5.10. The power spectrum of FSK can be displayed using the QT GUI frequency sink, as shown in Fig. 5.11. Since the constellation point for QPSK is '4', the constellation diagram for QPSK can be obtained as shown in Fig. 5.15 and for BPSK is '2' and so on.

The resultant waveform of BPSK for the given sample rate and the power spectrum are obtained, as shown in Fig. 5.12. The resultant constellation plot should be displayed only in two phases, as shown in Fig. 5.13.

The output for QPSK is similar to BPSK except for the constellation point, and the order of the Costas loop is mentioned as '4'. Thus the power spectrum and constellation plot can be obtained as shown in Figs. 5.14, 5.15 and 5.16.

For M-PSK, the value can be given for 'M' to obtain 8-PSK, 16-PSK, 64-PSK and so on, as shown in Figs. 5.17, 5.18, 5.19 and 5.20.

We obtain the constellation point as shown in Fig. 5.20 for the given appropriate constellation points in QAM block and the resultant power, and the decoded output is obtained as in Fig. 5.21 using FFT sink block in GNU radio.

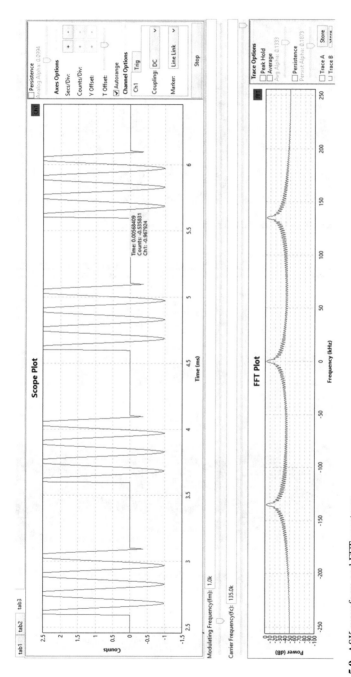

Fig. 5.9 ASK waveform and FFT spectrum

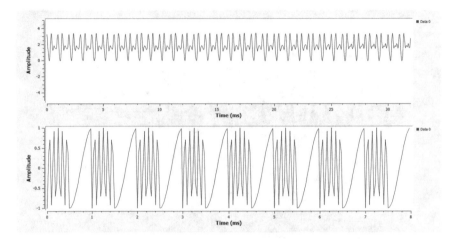

Fig. 5.10 FSK waveform before and after demodulation

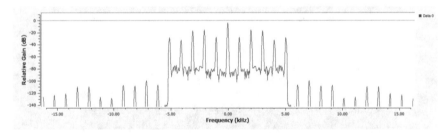

Fig. 5.11 FSK power spectrum

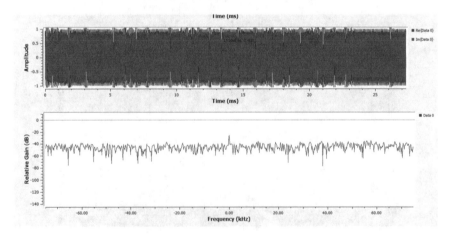

Fig. 5.12 BPSK waveform and power spectrum

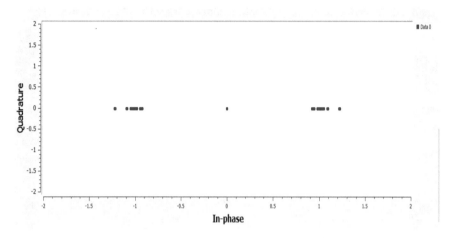

Fig. 5.13   BPSK constellation before and after demodulation

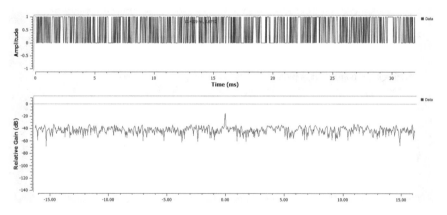

Fig. 5.14   QPSK demodulated output waveform and power

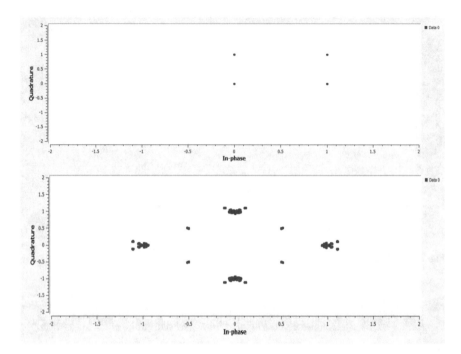

Fig. 5.15 QPSK constellation plot

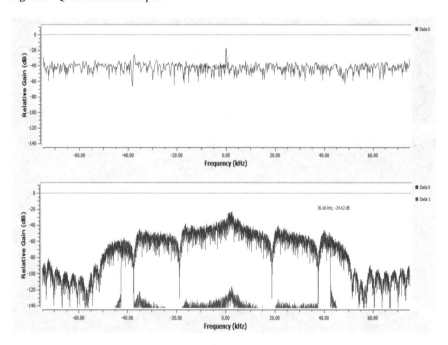

Fig. 5.16 Power spectrum of QPSK using bandpass filter

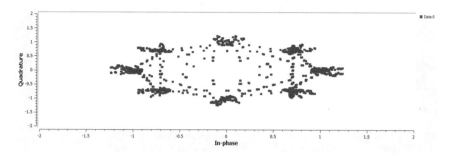

Fig. 5.17 Constellation plot for 8-PSK

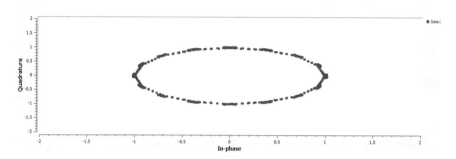

Fig. 5.18 Constellation plot for 16-PSK

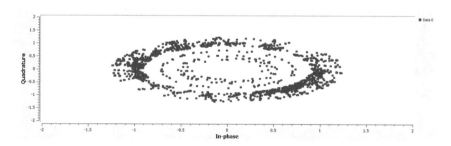

Fig. 5.19 Constellation plot for 64-PSK

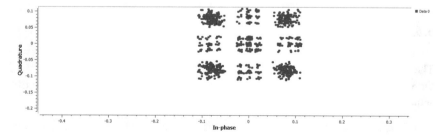

Fig. 5.20 16-QAM constellation plot

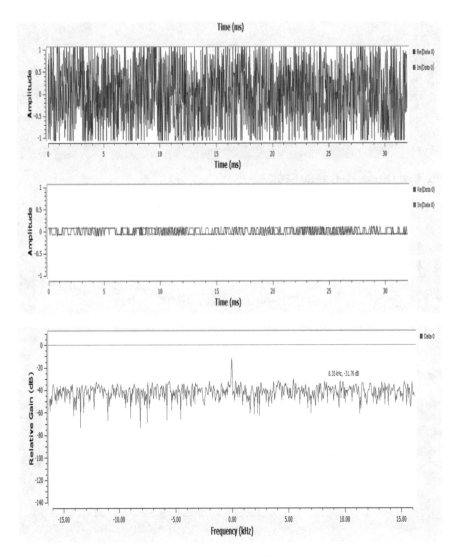

Fig. 5.21 QAM waveform before and after the noise and its power spectrum

5.6.2 Bit Error Rate for Digital Modulation Techniques

The BER is obtained for the best modulation schemes as shown in Fig. 5.22. QPSK is the most suitable modulation technique which can be adopted for the EPC sequence in an RFID system [14].

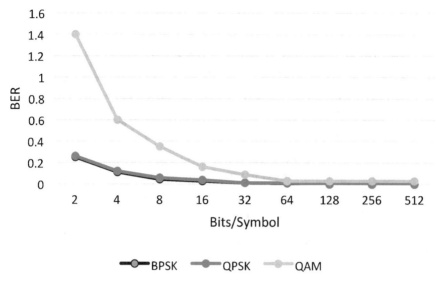

Fig. 5.22 BER vs Bits/Symbol

5.7 Conclusion

In this chapter, various modulation and demodulation schemes are discussed in detail and focused mainly on transmitting the EPC sequence with suitable modulation and demodulation scheme. We implemented in GNU radio, which is an adaptive SDR platform, and transmitted a random sequence using various modulation schemes and chose the suitable one with less BER and high SNR ratio. It is transmitted over AWGN channel model to analyse the differences in the received power. Thus, the power is analysed for UHF range of RFID frequencies.

References

1. S.C.S. Srinivas, A.S. Reddy, G.V. Sai Yeswanth, B.B. Jha, D.G. Kurup, EPC global Gen-2 RFID tag to reader communication simulation using GNU radio. Int. J. Adv. Res. Comput. Commun. Eng. **3**(4), 6314–6317 (2014)
2. M. Mitra, A random number generator for RFID tags. Int. J. Electron. Commun. Eng. Technol. **1**(1), 71–87 (2010)
3. A.T. Gaganpreet, H. Kaur, Implementation of file transfer with GNU-RADIO tool on SDR platform, in *International Conference on Soft Computing Applications in Wireless Communication – SCAWC 2017*
4. M.A. Hannan, M. Islam, S.A. Samad, A. Hussain, Modulation techniques for RFID transceiver using software defined radio. Int. J. Innov. Comp. Inf. Control **8**(10(A)), 6667–6692 (2012)
5. M. Mohaisen, H. Yoon and K. Chang, Radio Transmission Performance of EPCglobal Gen-2 RFID System, *2008 10th International Conference on Advanced Communication Technology*, Gangwon-Do, 2008, pp. 1423–1428. https://doi.org/10.1109/ICACT.2008.4494031

6. M. Mishra, A. Potnis, P. Dwivedy, S.K. Meena, Software defined radio based receivers using RTL – SDR: a review, in *Proceeding International Conference on Recent Innovations is Signal Processing and Embedded Systems (RISE-2017)*, 27–29 Oct 2017
7. S. Kuriakose, M. Jacob, QPSK modulation for DSSS-CDMA transmitter and receiver using FPGA. Int. J. Electron. Commun. Eng. Technol. **5**(12), 167–173 (2014)
8. L. Rebica, S. Rani, S. Kakkar, Performance analysis of various modulation techniques using GNU radio, in *International Journal of Computer Applications (0975 – 8887) International Conference on Advances in Emerging Technology (ICAET 2016)*
9. J.K. Lunagariya, K. Gokhruwala, K. Vachhani, Design analysis of digital modulation schemes with GNU radio, in *Second International Conference On Networks, Information & Communications,* May 2015
10. J. Muslimin, A.L. Asnawi, A.F. Ismail, A.Z. Jusoh, SDR-based transceiver of digital communication system using USRP and GNU radio, in *2016 International Conference on Computer & Communication Engineering,* 2016
11. M. Mukesh, L. Abhishek, R.R. Bhambare, QPSK modulator and demodulator using FPGA for SDR. Int. J. Eng. Res. Appl. **4**(4 (Version 1)), 394–397 (2014)
12. N. Kim, N. Kehtarnavaz, M. Torlak, LabVIEW-based software-defined radio: 4-QAM modem. Syst. Cybernet. Inform. **4**(3), 54–61 (2006)
13. N.S. Paujia, D. Astharini, O.N. Samijayani, SER and BER analysis using GNU radio for PSK and QAM modulation, in *International Seminar on Science and Technology Innovation,* Oct 2012
14. M.U. Singh, S. Kakkar, S. Rani, BER performance analysis of OFDM-MIMO system using GNU radio, in *MATEC Web of Conferences 57, 01022,* 2016
15. A. Zainudin, A. Sudarsono, I.G.P. Astawa, Reliability analysis of digital communication for various data types transmission using GNU radio and USRP, in *Industrial Electronic Seminar,* 2013
16. S. Hari Shankar Elango, E. Esakki Vignesvaran, Advanced ATM security system using RFID technology. Int. J. Appl. Eng. Res. **10**(20), 16887–16891 (2015)
17. X. Fan, W. Gong, J. Liu, I2tag: RFID mobility and activity identification through intelligent profiling. ACM Trans. Intell. Syst. Technol. **9**(1), 1 (2017). https://doi.org/10.1145/3035968
18. H. Ma, K. Wang, Fusion of RSS and phase shift using the Kalman filter for RFID tracking. IEEE Sensors J. **17**(11), 3551–3558 (2017). https://doi.org/10.1109/JSEN.2017.2696054
19. H. Ma, Y. Wang, K. Wang, Automatic detection of false positive RFID readings using machine learning algorithms. Expert Syst. Appl. **91**, 442–451 (2018). https://doi.org/10.1016/j.eswa.2017.09.021
20. M. Huerta, J. Ferreira, L. Rodriguez, R. Clotet, R. Gonzalez, D. Rivas, Design of a building security system in a university campus using RFID technology. Paper presented at the *2017 IEEE 37th Central America and Panama Convention, CONCAPAN 2017*, 1–6 Jan 2018, https://doi.org/10.1109/CONCAPAN.2017.8278525
21. K. Fan, W. Jiang, H. Li, Y. Yang, Lightweight RFID protocol for medical privacy protection in IoT. IEEE Trans. Ind. Inform. **14**(4), 1656–1665 (2018). https://doi.org/10.1109/TII.2018.2794996
22. S. Anandhi, R. Anitha, V. Sureshkumar, IoT enabled RFID authentication and secure object tracking system for smart logistics. Wirel. Pers. Commun. **104**(2), 543–560 (2019). https://doi.org/10.1007/s11277-018-6033-6
23. P. Arulmozhi, J.B.B. Rayappan, P. Raj, A lightweight memory-based protocol authentication using radio frequency identification (RFID) (2019). https://doi.org/10.1007/978-981-13-1882-5_14
24. S. Lee, J. Kim, N. Moon, Random forest and WiFi fingerprint-based indoor location recognition system using smart watch. HCIS **9**(1) (2019). https://doi.org/10.1186/s13673-019-0168-7
25. J.V. Gorabal, Novel implementation of multi modal biometric approaches to handle privacy and security issues of RFID tag. IIOAB J. **7**(3), 60–65 (2016)
26. M.A. Ferrag, L. Maglaras, A. Argyriou, D. Kosmanos, H. Janicke, Security for 4G and 5G cellular networks: A survey of existing authentication and privacy-preserving schemes. J. Netw. Comput. Appl. **101**, 55–82 (2018). https://doi.org/10.1016/j.jnca.2017.10.017

Chapter 6
Design of Classifiers

S. Padma

6.1 Introduction

Data mining can be defined as mining of data from data warehouse. It can also be said as mining essential knowledge or knowledge discovery from data (KDD). Of all the various data mining techniques, classification is an important one. Classification is a major technique in data mining where a group of data is divided and given categories for various analyses and effective prediction. The vital goal of classification is to find the category/class of a new data which enters into the group. For both structured and unstructured data, classification can be done. Categorizing the data into suitable classes based on the available features is said to be classification. The prediction and classification of data is done based on the training of the categorized data. By learning the characteristics of the data in training, the classifier will classify the input while testing process takes place.

Classifiers like decision trees, nearest neighbor, naïve Bayes, logistic regression, and neural networks perform well. The applications of classifiers are becoming broader day by day. To deal with several new applications and large amount of data, neural network concepts perform better compared to other methods. Being inspired by the biological nervous system of the human brain for processing information, neural network concept became popular. The working of the concept is by the elements (neurons) which are strongly interconnected with various applications of artificial intelligence; machine learning is an important one which has the ability to learn automatically and gain experience without pre-programming. Machine learning concentrates on the data that is to be accessed and made the systems learn themselves.

S. Padma (✉)
Department of Computer Applications, K.S. Rangasamy College of Arts and Science
(Autonomous), Tiruchengode, Tamil Nadu, India

© Springer Nature Switzerland AG 2019
G.R. Sinha (ed.), *Advances in Biometrics*,
https://doi.org/10.1007/978-3-030-30436-2_6

Various cognitive methods are employed in the machine learning concepts to acquire more knowledge and discovery techniques. Learning can be done either supervised (with teacher) or unsupervised (without teacher). While incorporating the cognitive tasks in learning methods, the results are highly considerable. The learning may be either by inductive or deductive or incremental or non-incremental. Several studies related to cognition clearly disseminate that cognitive task inclusion in algorithms finds solutions to several problems. The theory extracts the needy, related elements to solve the problem and thereby improves generalization.

6.2 Cognitive Principles

"Metacognition" can be defined as "thinking about thinking." But it is not simple as its definition. Many educational psychologists are under research on the metacognitive experiences for the past two decades. Still there is a thirst on the word metacognition. The main researches on metacognition focuses on mind theory (developmental psychology), on meta-memory (experimental psychology), and on self-regulated learning (educational psychology).

As per the definition of Nelson, metacognition is said to be prototype of cognition located at the meta-level. Cognition is highlighted in the object level of metacognition. From the above statements, it is clear that at the meta-level the metacognition functions, whereas the prototype of the cognition is metacognition. Both the cognition and metacognition are linked through the control and monitoring functions.

One more model referred by Nelson and Narens on metacognition is given in Fig. 6.1. It also suggested about the control and monitoring process. The above-said model contains two divisions cognitive and metacognitive component. The cognitive division has two interconnected steps called the meta-level and object-level. The directions of the information flow are the metacognitive components. The flow of information is controlled and monitored by two signals, namely, control and monitoring signal. The cognitive component is in the object level which includes concentration, learning, processing of languages, solving various problems, etc. The meta-level includes the cognitive component, learning constraints, goals to be achieved, and several strategies in developing knowledge. The object-level contents are monitored by passing the signals to the meta-level. The meta-level contents are controlled through the signals to the object-level. The mistakes at the object-level are taken care by monitoring signals. The mistakes at the meta-level are taken care by the control signals. This simple method of learning through the available skills is given in the model.

Cognitive model deals with memory. In addition the metacognitive knowledge includes the cognitive content together with the language, memory storage, etc. It also includes the information about a person, product, different views of a content, and more. Dealing with humans, the metacognitive knowledge describes the actions and reactions on various jobs. The actions of humans, relationships between their activities, and processing of thinking come under the metacognitive knowledge.

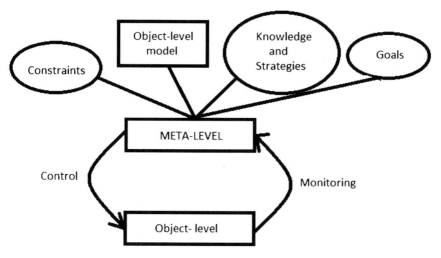

Fig. 6.1 Nelson-Narens model

On the whole the metacognitive knowledge speaks well about "knowledge about knowledge" on various situations, the decision-making, the experiences, and other related themes. The enrichment of metacognitive knowledge happens continuously. It integrates the information and updates the necessary data; clear differentiation of data will be done. The above said are monitored by the cognitive component and controlled by measures. The update is done through experiences, awareness in the job, communication, and other interactive modes.

6.3   Classification Problem

Let us take the values$\{(x1,y1), (x2,y2), \ldots (xt,yt), \ldots\}$ where $x_t \in \Re^m$ is the observed features of m-dimension and $x_t \in \Re^m$ is its class identifier, where n is the aggregate number of groups. The classification is represented using the values 1 and -1. Once the observed feature x belongs to class label c of the value set, assign it to 1 or else to -1.

$$y_j = \begin{cases} 1 & \text{if } j == c \\ -1 & \text{otherwise} \end{cases}, \quad j == 1, 2, \ldots, n$$

The principle of the classifier is to predict the corrected class label with nominal accuracy with the available input random variables along with knowledge and x. The new identification of x is based on the prediction of coded class label y which is said to be as classification. Thus it estimates the relationship within the feature and the coded class label.

6.4 Classifier Models

Neural networks capture data through batch learning or online/sequential learning mode. In batch learning the inputs are sent into the networks chunk by chunk. For some applications the entire training samples may not be received in advance. At this situation batch learning classification algorithm performance accuracy is bad. This can be overcome by the sequential learning algorithms where the data is sent into the network sequentially. Below are the few learning algorithms that work based on the cognitive principles of the human brain.

The normal execution of the sequential learning methods read the training data sequentially and read only once in its learning period. The architecture of this type of algorithms is constructive and fixed. The generalization of the classifier will be poor if the training sample contains similar data which also leads to overfitting. The performance of the classifier gets affected by the training samples also. The approximations of the algorithms are highly influenced by the control parameters. Many of the sequential algorithms have the drawback in giving good results as the sequence of the input sample is to be altered for better performance.

6.5 Self-Regulatory Resource Allocation Network (SRAN)

The SRAN classifier comes under the sequential learning methods. It is suitable for both binary and multicategory classification-controlled mechanism known as self-regulatory learning. SRAN uses the radial basis function (RBF) as a main block. Several groups of control parameters are combined with the RBF, to execute the working of the network under the concept of self-controlled learning. Of all the various activation functions in RBF networks, Gaussian basis function is the most commonly used function. The learning method of sequential learning algorithm is it reads the input data sequentially and only once. If the training is on the same data for a long time, it will affect the classifier performance. The classifier performance is controlled by several parameters associated with the algorithms. The traditional RBF architecture is modified based on the self-regulatory parameters. Figure 6.2 clearly indicates the work of SRAN.

The working principles of self-regulation system are deleting samples based on criteria, growth of network, and updation of parameters.

- Sample delete condition: In order to avoid overtraining of data, check before deleting the sample that the absolute maximum error is less than 0.05.
- Network growth criteria: If the maximum error is within the threshold value and the class is not equal to the predicted class, then the neurons can be added to the network.
- Updation of parameter: Extended Kalman filter (EKF) is used for parameter updation if and only if the actual and the predicted classes are the same. It also checks if the maximum error value is above the threshold error.

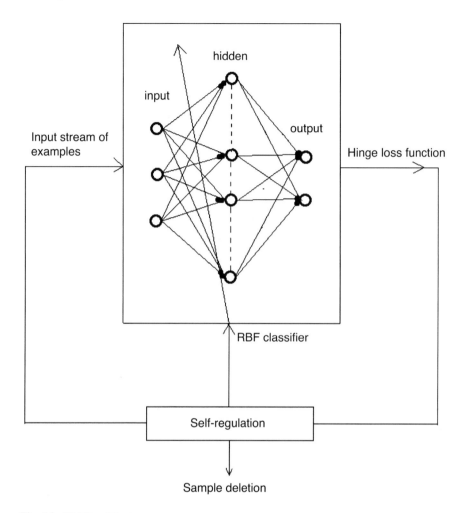

Fig. 6.2 SRAN architecture

6.6 Metacognitive Neural Network (McNN)

Inspired by the cognitive mechanisms of the human brain together with the self-controlled learning paved the way to develop. McNN works as per the method delivered by Nelson and Narens given in Fig. 6.1. The architecture comprises dual parts. The foremost one is the cognitive part and the latter the metacognitive part. Information about the samples moves from the cognitive part to the metacognitive part which was said to be monitoring. Flow of information in the opposite direction is said to be control. The traditional three-layered RBF network is the cognitive component. Metacognitive component comprises the copy of the cognitive component combined with the knowledge measures and learning strategies. The

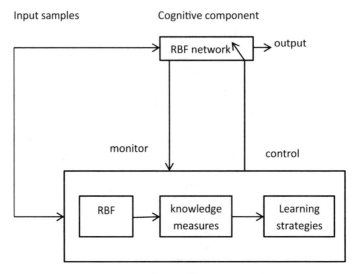

Fig. 6.3 McNN architecture

knowledge measures are estimating the class label, maximum hinge error, classifier confidence, and class-wise significance. The learning strategies are sample delete, parameter update, sample reserve, and neuron growth. Whenever a new sample arrives, the sample executes based on the principles of what to learn, when to learn, and how to learn. Then McNN architecture (Fig. 6.3) is as follows.

6.6.1 Learning Strategies

To enhance the learning of the sequential learning algorithm, the concept of self-regulated learning derived from the human learning principles is used. The principles of human learning that influences machine learning in the method are what to learn, when to learn, and how to learn along with several learning strategies. The metacognitive part of the McNN selects any one of the below-defined strategy when new training sample arrives.

- Sample delete condition: Delete the training data that arrives to the cognitive component if its knowledge is already in it. Learning of that sample is not required.
- Neuron growth strategy: Add new neurons to the network if the class is not in the list and the error is within the threshold value.
- Parameter update strategy: To improve the generalization and reduction of error, EKF is used for parameter update.

- Sample reserve method: Not all the training data are meant for learning unless they are significant. If not so they can be reserved for future reference to retrieve some additional information that helps for better classification. If a sample does not suit to the above category, then it can be removed.

6.6.2 Knowledge Measures

- Estimated class label: Estimation of class label based on the predicted output is obtained.
- Maximum hinge error: Most of the classifiers were developed by incorporating mean square error method to calculate error that has less accuracy when compared to the hinge loss function which is used in some classifiers.
- Class-wise significance: Performance of the classifier depends on the class-wise distribution.

6.7 Metacognitive Fuzzy Inference System (McFIS)

Described by Nelson and Narens method of metacognition, McFIS contains two divisions, namely, a cognitive and the metacognitive. The cognitive part comprises the TSK-type-0 neuro-fuzzy inference system. Self-controlled learning of the cognitive component leads to the metacognitive component. The learning method checks the network knowledge based on the past and the present samples. The network is controlled by the principles what to learn, how to learn, and when to learn. Measuring the knowledge of the network and measuring the errors of the samples are utilized as monitory signals. The monitory signals are used to select the learning method suitable for the knowledge available in the current sample. After this selection the samples are added to the fuzzy inference system (FIS). The architecture of the McFIS (Fig. 6.4) comprises four layers, namely, the input layer which holds input samples, second layer which is the Gaussian layer where the activation function adds weights to the inputs, third layer in which the values of the neurons are normalized, and the fourth layer where the output is retrieved.

The metacognitive component includes the copy of the cognitive component along with the monitory and control signals. Instead of mean square error, hinge loss error function is used to measure sample error. As monitory signals knowledge-based measures are used. The learning measures are executed with the help of the control signals using strategies like sample delete, sample learning and sample reserve.

The above-said conditions are as follows:

- *Sample deletion strategy*: Avoids overtraining and reduces computational effort by preventing the sample learning recursively.

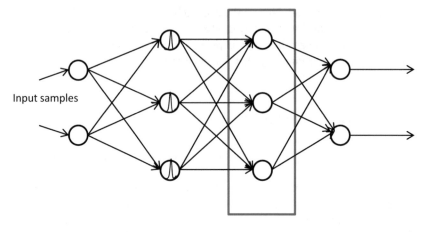

Input samples

Input layer   Gaussian layer   normalization layer   output layer

Fig. 6.4 McFIS cognitive component architecture

- *Sample learning strategy*: Allocation of new rule ("growing rule"), rules for updating parameters ("updation of parameter"), or deletion of same sample ("rule pruning") which represents the how-to-learn principle of metacognition.
- *Sample reserve strategy*: Helps in fulfilling one of the metacognitive tasks – what to learn – by deleting the samples that have no novel information. How-to-learn principle is fulfilled by the sample reserve strategy. Learning immediately or later fulfills the metacognition principle when to learn.

6.8   Projection-Based Learning with McNN (PBL McNN)

McNN has dual components. A traditional RBF network with single hidden layer is the cognitive component, whereas the metacognitive component contains the model of the cognitive component along with knowledge measures and learning strategies. The cognitive component of the McNN learns from stream of input data. Through the input neuron, hidden neurons are grown to perform a network structure. To approximate the classifier performance, the weights of the hidden neurons are updated. The parameters of the Gaussian function are updated when the hidden neurons are added to the network. The calculation of the Gaussian function is mainly based on the centers and width of the current samples. The weights of the neurons are calculated using the projection-based learning method. It works to convert the set of linear problem into linear equations to minimize the error through optimal weight.

Radial basis function neural network makes a nonlinear mapping from input space R^n to the output space R^m. R^n is an input stream that is denoted by x_i (for

$i = 1, 2, 3, \ldots n$) and R^m is output vector space that is denoted by y_i (for $i = 1, 2, \ldots m$). The jth hidden neuron of the radial basis function, which is computes a Gaussian function as below

$$Z_i(x) = \exp\left(-\frac{\|x - c_j\|}{2\sigma_i^2}\right) \quad j = 1, 2, \ldots m \tag{6.1}$$

where x is input feature vector with n dimension, c_j is the center of Gaussian vector of i, and σ_i is width of the hidden layer.

The width of the hidden layer σ_i is calculated by

$$\sigma_j = \sqrt{\frac{1}{m_j} \sum_{i=1}^{m_j} d^2\left(c_j - x_i\right)} \tag{6.2}$$

When a sample is used to update the output weight parameter by projection-based learning algorithm then

$$\frac{\partial J\left(W_K^t\right)}{\partial w_{pj}} = \frac{\partial J\left(W_K^t\right)}{\partial w_{pj}} + \frac{\partial J_t\left(W_K^t\right)}{\partial w_{pj}} = 0, \quad p = 1, \ldots, K; \quad j = 1, \ldots, n \tag{6.3}$$

With respect to zero, equating first partial derivative and re-arranging (6.3) gives as follows,

$$\left(A^{t-1} + \left(h^t\right)^T W_k^t - \left(B^{t-1} + \left(h^t\right)^T \left(y^t\right)^T\right) = 0 \tag{6.4}$$

By substituting $B^{t-1} = A^{t-1} W_K^{t-1} A^{t-1} + \left(h^t\right)^T h^t = A^t$ and adding or subtracting the term $\left(h^t\right)^T h^t W_K^{t-1}$ on both sides, Eq. (6.4) is reduced to

$$W_K^t = \left(A^t\right)^{-1} \left(A^t W_K^{t-1} + \left(h^t\right)^T \left(y^t\right)^T - h^t W_K^{t-1}\right) \tag{6.5}$$

In conclusion, the output weight can be updated as

$$W_K^t = W_K^{t-1} + \left(A^t\right)^{-1} \left(h^t\right)^T \left(e^t\right)^T \tag{6.6}$$

This study is to utilize the hinge loss error function instead of mean square error to estimate the error rate between actual and predicted data that reduces the energy function. Projection-based learning algorithms convert linear problems into set of linear equation. The linear equation solves the classification problem with optimal weights and minimizes the energy function.

6.9   Metacognitive Extreme Learning Machine (McELM)

The aforementioned classifiers work on the principles of sequential learning algorithms. In the sequential learning methods, the data are learned one by one and using the self-regulatory concept. The self-regulatory concept prevents the neurons from overtraining which increases the generalization of the classifier. Extreme learning machine (ELM) is a fast learning method which is used to solve regression as well as classification problems. ELM networks are computationally intensive and train the data within a short duration. Input parameters are chosen randomly, whereas the output weights are estimated analytically. ELM is a batch learning process where the data are sent to the network chunk by chunk and the network is fixed a priori.

The metacognitive principles of Nelson-Narens model have proved its excellence in various fields. The human learning principles when included with the ELM working method give a better classifier. ELM with three layers, namely, input neurons, hidden neurons, and output neurons, forms the cognitive component of McELM. Hidden layer uses the q-Gaussian radial basis function, while the input and output layer neurons of McELM are linear.

The metacognitive component uses the self-regulatory mechanism along with the learning strategies and knowledge measures. The learning strategies are sample deletion, sample learning, and sample reserve. Knowledge measures are predicted class label and maximum hinge loss error. The training performance of the classifier along with the training speed is best when compared to other sequential learning algorithms.

6.10   Summary

The learning principles of the human brain when inculcated into the learning principles of neural networks produce better results. The above-said sequential learning algorithms proved its efficiency in training and also in time duration. Classifiers are intended to use the self-regulatory mechanism; thereby the neural network executes the neurons same as the human brain neurons. The cognitive principles when incorporated into the classifier with the learning strategies and knowledge measures work effectively.

Bibliography

1. G.P. Zhang, Neural networks for classification: a survey. IEEE Trans. Syst. Man Cybern. Part C Appl. Rev. **30**, 451–462 (2000)
2. S. Sivanandam, S. Deepa, *Introduction to Neural Networks Using Matlab 6.0* (Tata McGraw-Hill Education, New Delhi, 2006)

3. S. Sivanandam, S. Deepa, *Principles of soft computing (with CD)* (John Wiley & Sons, India, 2007)
4. K. Tyagi, X. Cai, M.T. Manry, Fuzzy C-means clustering based construction and training for second order RBF network, fuzzy systems (FUZZ), in *2011 IEEE International Conference on IEEE* (2011), pp. 248–255
5. I. Czarnowski, P. Jędrzejowicz, Kernel-based fuzzy C-means clustering algorithm for RBF network initialization, in *Intelligent Decision Technologies*, (Springer, Italy, 2016), pp. 337–347
6. G. Kayhan, A.E. Ozdemir, İ. Eminoglu, Reviewing and designing pre-processing units for RBF networks: initial structure identification and coarse-tuning of free parameters. Neural Comput. Applic. **22**, 1655–1666 (2013)
7. H. Sarimveis, P. Doganis, A. Alexandridis, A classification technique based on radial basis function neural networks. Adv. Eng. Softw. **37**, 218–221 (2006)
8. E.A. Lim, Z. Zainuddin, An improved fast training algorithm for RBF networks using symmetry-based fuzzy C-means clustering. Matematika **24**, 141–148 (2008)
9. H.-G. Byun, An identification technique based on adaptive radial basis function network for an electronic odor sensing system. J. Sens. Scie. Technol. **20**, 151–155 (2011)
10. A. Esmaeili, N. Mozayani, Adjusting the Parameters of Radial Basis Function Networks Using Particle Swarm Optimization, in *2009 IEEE International Conference on Computational Intelligence for Measurement Systems and Applications* (IEEE, 2009), pp. 179–181
11. S. Mitra, Fuzzy radial basis function network: a parallel design. Neural Comput. Applic. **13**, 261 (2004)
12. S. Mitra, J. Basak, FRBF: a fuzzy radial basis function network. Neural Comput. Applic. **10**, 244–252 (2001)
13. S.-B. Roh, S.-K. Oh, Identification of plastic wastes by using fuzzy radial basis function neural networks classifier with conditional fuzzy C-means clustering. J. Elect. Engg. Tech. **11**, 1921–1928 (2016)
14. A. Kaushik, A. Soni, R. Soni, Radial basis function network using intuitionistic fuzzy C means for software cost estimation. Int. J. Comput. Appl. Technol. **47**, 86–95 (2013)
15. I. Czarnowski, P. Jędrzejowicz, *An Approach to RBF Initialization with Feature Selection, Intelligent Systems* (Springer, Italy, 2015b, 2014), pp. 671–682
16. W. Kaminski, P. Strumillo, Kernel orthonormalization in radial basis function neural networks. IEEE Trans. Neural Netw. **8**, 1177–1183 (1997)
17. M. Awad, H. Pomares, I.R. Ruiz, O. Salameh, M. Hamdon, Prediction of time series using RBF neural networks: a new approach of clustering. Int. Arab J. Inf. Technol. **6**, 138–143 (2009)
18. I. Czarnowski, P. Jędrzejowicz, An approach to RBF initialization with feature selection, in *Intelligent Systems'2014: Proceedings of the 7th IEEE International Conference Intelligent Systems IS'2014*, September 24–26, 2014, Warsaw, Poland,, Volume 1: Mathematical Foundations, Theory, Analyses, ed. by P. Angelov, K. T. Atanassov, L. Doukovska, M. Hadjiski, V. Jotsov, J. Kacprzyk, N. Kasabov, S. Sotirov, E. Szmidt, S. Zadrożny, (Springer International Publishing, Cham, 2015), pp. 671–682
19. W. Jia, D. Zhao, T. Shen, C. Su, C. Hu, Y. Zhao, A new optimized GA-RBF neural network algorithm. Comput. Intell. Neurosci. **2014**, 44 (2014)
20. C.-L. Lin, J. Wang, C.-Y. Chen, C.-W. Chen, C. Yen, Improving the generalization performance of RBF neural networks using a linear regression technique. Expert Syst. Appl. **36**, 12049–12053 (2009)
21. G.S Babu, R. Savitha, S. Suresh, A projection based learning in meta-cognitive radial basis function network for classification problems, in *The 2012 International Joint Conference on Neural Networks (IJCNN)* (IEEE, 2012), pp. 1–8
22. G.S. Babu, S. Suresh, Meta-cognitive RBF network and its projection based learning algorithm for classification problems. Appl. Soft Comput. **13**, 654–666 (2013)
23. G.S. Babu, S. Suresh, Sequential projection-based metacognitive learning in a radial basis function network for classification problems. IEEE Trans. Neural Netw. Learn. Syst. **24**, 194–206 (2013)

24. K. Subramanian, S. Suresh, R.P. Cheng, A modified projection based learning algorithm for a meta-cognitive radial basis function classifier, in *Cognitive Computing and Information Processing (CCIP), 2015 International Conference on IEEE*, (2015), pp. 1–6
25. J.C. Bezdek, R. Ehrlich, W. Full, FCM: The fuzzy c-means clustering algorithm. Comput. Geosci. **10**, 191–203 (1984)
26. S. Suresh, N. Sundararajan, P. Saratchandran, Risk-sensitive loss functions for sparse multi-category classification problems. Inf. Sci. **178**, 2621–2638 (2008)

Chapter 7
Social Impact of Biometric Technology: Myth and Implications of Biometrics: Issues and Challenges

Kavita Thakur and Prafulla Vyas

7.1 Introduction

However biometric system has versatile uses among various fields for identification and verification but it faces some challenges and issues. This chapter deals with the myth and implication of biometrics as well as various challenges and issues which hinder the implementation, usability, and adoptablity of biometric system. Biometrics is the better authentication mode to identify persons. Basically, biometrics indicates what you are rather than what you have, i.e., possession, such as keys, passport, smartcard, etc., or what exactly you remember in mind such as secret codes, PIN codes, etc [3]. Every person has their own unique biometrics which can't be shared, stolen, or even forgotten. Clearly any solid individual recognizable proof ought to incorporate biometrics in light of the fact that individuals don't lose their physiological and organic attributes. The biometric system explores various physical or biological traits like finger veins, face, retina, signature, voice, etc. to identify the individuals. Identification system is widely used by most software development organization to trace out their employee time and attendance verification. Biometric system tries to avoid proxy identification which is very critical in time-bound organization. It saves a lot of remuneration paid to employees. According to the US Department of Commerce, the business of country bears the cost of $50 billion annually because of employee time theft and insincerity. Basically, biometric system eliminates time theft and ghost employees, increases productivity, reduces administrative costs, as well as reduces payroll errors. As compared to biometric

K. Thakur (✉)
School of Studies in Electronics and Photonics, Pt Ravishankar Shukla University,
Raipur, CG, India

P. Vyas
Disha College, Raipur, CG, India

© Springer Nature Switzerland AG 2019
G.R. Sinha (ed.), *Advances in Biometrics*,
https://doi.org/10.1007/978-3-030-30436-2_7

system, traditional surveillance systems like barcode swipe cards, cameras, personal identification number, etc. are less effective solutions. It is a cost-effective solution to identify persons. There are minimal social issues with respect to the security and comfort of utilizing biometric framework. However, there is debate whether these are safe and suitable to use. The government still tries to improve the system to make sure food, medicines, fuel, and other assistance reach the needy persons. Judicial laws ruled that government's biometric ID system for citizens is legal. The government has begun to pass regulations to regulate data collection, protection, and sharing with others. Biometric identifications are harder to steal than other traditional traits.

It has been conceded to by the administration and the business that biometric acknowledgment is presently turning into an unavoidable truth. In any case, such developing innovations have various myths, issues, misunderstandings, and challenges.

Biometric science which arrangements the ID or check of an individual ward is dependent on its own physiological and/or social attributes faces lot of myths. There are a number of issues and legal considerations that have been considered as regards the full-fledged implementation of biometric system. In recent times, awareness of biometrics for common man was limited to its use in covert operative spine chillers or dread ingraining instruments of condition of corporate observation in theoretical fiction, forensic or as investigative tools; and was supposed to be ultimate in areas where it was applied. There has been a sea change in areas where biometrics can be applied. From around 2011, biometric advances seem stately for more extensive use. The nationalized security and follow-ups of individuals who possess passports, visa, etc. to cross the border are big issues to ensuring authorized entry. Every one of the personalities and fringe crossing records must be connected to biometric information. Every person has their exclusive features which are almost stable throughout life and are induced into advanced biometrics to distinguish them from another person. It minimizes fake representation and is implemented to provide user-friendly operating conditions [2].

7.2 Biometric Myths and Misrepresentation

Biometrics is an undoubtedly emerging technology than other technologies. In any case, most quite, it is executed in an alliance where they are censured and misjudged more than different advances. Here, those legends are investigated and clarified why they endure and why these discernments aren't right. Biometric system is versatile which is suitable for all applications.

Someone thought that single biometrics is appropriately fit to all applications. There are a lot of versatile applications, namely, authentication of driver license, border control, access control, etc. which cannot be fulfilled by single biometrics. Various important factors like client conditions, size of client populace, society acknowledgment, framework cost, and so forth have to be considered to select biometrics for an appropriate application. Adequacy of a biometric innovation relies upon the how and where it is utilized. Each biometric innovation has its own qualities and shortcomings that ought to be assessed in connection to the application before usage. Biometric tools cannot be generalized in terms of usages. It can give results/identify only the particular which are saved within.

Particular Biometrics Is Unique for Each Individual
The facts confirm that biometric qualities of every individual are interesting if and just whenever broken down with adequate detail. However, this statement is not fully true because of some limitations as well as intrinsic intraindividual varieties after some time. Generally, identical biometric traits of any two individuals are not common; it is likewise amazingly impossible that two estimations of a similar individual would give indistinguishable representation. Therefore, resilience ought to be a significant component of a useful biometric framework to beat such issues, permitting coordinating of biometrics regardless of estimation commotion and fleeting variety. System accuracy is quantifying by single number sample.

It is difficult to obtain a precise decision easily by comparing and matching single sample. The error rates which are just numbers ought to be translated with regard to the biometric innovation being used alongside the current application.

Biometric System Is Preconfigured
Biometric system depends upon the database of biometric templates which have been extricated from a biometric test. It does not immediately get user-friendly. Captured and stored features which could be restricted as far as size and genuine assortment are utilized to prepare the framework. It is required to adjust to the real working condition and is to be tuned to the detecting gadgets and securing conditions.

Genuine Exactness Execution Can Be Anticipated in the Design Phase
Real accuracy prediction means that a particular system will provide real accuracy performance by actual population. It can't be anticipated in the plan stage; it very well may be only evaluated.

Biometrics Is Costlier Than Other Surveillance System
Biometrics can be costly, yet not generally. It relies upon which biometrics is utilized, how they are utilized, and the size of the execution. As per the biometric technology, the camera of iris recognition is generally more costly than the fingerprint recognition system; obviously it relies upon the manufacturer, with expenses for a similar sort of innovation fluctuating broadly. For instance, the cost of biometric modules depends upon the area where it is utilized. For example, the cost of iris camera that is utilized to control access at an assembly plant is much impressive as compared to the cost of the iris camera that is utilized in a region of highly sensitive or of high throughput, viz., in airplane terminals. In this manner, biometrics that are utilized is generally safe purchaser tools, similar to PC incorporated fingerprint reader and face acknowledgment empowered. Smart telephones possess inexpensive biometric traits than the utilization of biometrics for controlling site access at an administration guard site. It is an essential on the

grounds that, and as we see on numerous occasions with different innovations, there are top-of-the-line and low-end applications, with changing degrees of speed and dependability. Biometrics that must be quick or potentially utilized in zones of high hazard will dependably be far costlier than those utilized in generally safe/low-throughput conditions. Scale and unpredictability likewise have an impact in the expense to actualize.

Multiple Biometrics Perform Better Than Single Biometrics
Instinctively, utilizing different biometrics per individual and in addition utilizing numerous examples per biometrics may be valuable; anyway there is no assurance this really beats individual biometrics which may force much overhead with no huge effect to the framework execution.

Decision Threshold Is Not Utilized by Biometric Framework
Any biometric verification framework ought to incorporate a decision stage to choose whether the approaching individual is acknowledged or not. Such decision is chiefly reliant on the consequence of highlight matcher, so how such decision could be taken without characterizing a limit to show the acknowledgment dimension of coordinating. To build the security of the framework, the limit can be expanded, which diminishes false accept mistakes and expands false reject errors.

Extracted Components Can Be Utilized with any Match Engine
Various highlights may have basic portrayal, yet they are distinctive as far as their physical significance and their inclination are concerned. A coordinating engine is fundamentally worried about estimating the separation between highlights with regard to their component spaces. So various highlights taken from the various components space will naturally force utilizing distinctive coordinating engine, regardless of whether clearly includes have basic portrayal and various matchers have basic base (estimating separation).

Accuracy of Biometrics Is Defined by the Number of Templates
It is a misconception about the accuracy of biometric system that large templates provide better accuracy. System security is not exactly dependent on the biometric template. Additionally, vast templates may crumble framework execution, since having substantial layout implies progressively exact subtleties for an individual; this may influence the framework resilience to biometric variety for a person. Along these lines, huge layouts are not really legitimate [16, 18].

Biometrics Implies Complete Protection
No framework can be completely impeccable, particularly while considering the assaults of expert programmers. Figure 7.1 indicates phases of confirmation framework and enlistment framework and purposes of conceivable assault in a conventional biometric framework. No innovation gives 100% certification, and the present condition of face recognition is still a long way from being totally developed. After rigorous testing of face recognition at US airport and open spot, it is observed that face recognition system still needs a lot of advancements.

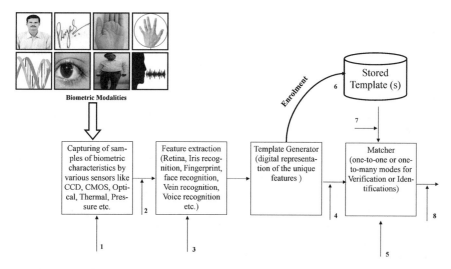

Fig. 7.1 Susceptibility points of a Biometric System

Biometric Frameworks Attack Our Confidentiality

Numerous court discoveries infer that the utilization of biometrics does not attack a person's affable freedoms or security, albeit individual perspectives are abstract and may contrast. A very much ideal biometric framework execution ought to consider those issues in the planning stage.

The Best Error Rates Have the Most Exact Framework

In general, data which are used to evaluate system might be gathered in very controlled environment and improbable. It may mislead the accuracy result which is highlighted by vendors. So, it is better to evaluate different systems on standard open space database to build up reasonable execution assessment in a realistic way. In addition, regardless of whether standard databases are utilized in the assessment stage, there is no assurance that such reports were for the whole set; it may be accounted for a subset of the information or without following a standard testing convention.

Biometric Sensors Are Not Sterile

Some people consider severely hygienic issues of biometrics. A large portion of biometric securing gadgets need direct contact. Most reported health issues are actually like those experienced in everyday life such as touching a doorknob, touch pad, and screen to interact with ATM machine. The scanners which scan the retina and iris don't sparkle lasers into the eye and thereby are not harmful. And many biometric implementations nowadays use touchless sensors for physical and behavioral trait recognition, eliminating this concern entirely.

Biometrics Is Excessively Mind-Boggling

It is undoubtedly accepted that hidden calculations utilized in biometric innovation are unpredictable and hard to get. As a customer we also don't know the internal mechanism of mostly used appliances or devices like microwave or car. However we can most conveniently operate them. It is to be understood by the biometric facilitators that complex circuitry idea must be kept away from the ordinary user. Every manufacturer prepares the document to simplify biometric operation.

Your worry ought to be with the multifaceted nature of the interface – both of the biometric gadget itself and the enrolment innovation. Once more, great producers will structure an interface that is easy to work by individuals of shifting specialized inclination. It is basic that any biometric gadget can be utilized by individuals who have practically zero information of innovation. The key is knowing how biometrics can function in your specific association and how you will profit, regardless of whether that is through improved proficiency, lower overheads, decreased introduction to hazard, or a mix of the abovementioned. The intricacy generally comes when building up how to coordinate the innovation. However, similarly as with some other business frameworks, e.g., HR, T&A, or CRM, this is generally reliant on what you as of now have and how you need biometrics to fit inside that.

Biometrics Violate Confidentiality

Biometrics can possibly disregard security and contradict laws like the Information Assurance Act if it is in the wrong hands. However, that is the key term here: in the wrong hands. Any innovation, when utilized improperly, regardless of whether a customer relationship management framework or a database utilized for holding individual financial data, can be a danger to our security. What's more, that is actually why we have laws like the Information Assurance Act to direct how we handle and use information.

Processing of Biometrics Is Tedious and Time-Consuming

Basically processing of biometrics is evaluated by its speed. Several factors are responsible to affect the speed of a biometric gadget. The speed depends upon the inquiry and matches the database. Templates of iris recognition are lighter, so they set aside less effort to coordinate, not at all like some unique mark formats, which are commonly vast and awkward, which makes coordinating more time-consuming. The dimension of association required with the client. For instance, some face acknowledgement advances require no physical communication by any stretch of the imagination, which is the reason they are regularly utilized in regions of extensive throughput, as airport. It is highly recommended to visit a site to perceive how the innovation is utilized and along these lines the speed at which individuals are handled.

A Fingerprint Faces Lots of Security Constraints

Most people think that scanned fingerprint data which is stored in biometrics is not safe. It may be used illegally or misused. But it is totally a misconception; its scanning method is drastically different from traditional fingerprint scanning methods. It doesn't store the solid scanning image directly; it just identifies and compares it using mathematical algorithm. There is some dispute that stored data can be used to regenerate your fingerprint images. Biometrics just decodes the information whenever your finger is put on the scanner.

A Biometrics Is Only Appropriate for a Large Firm
Generally, employee's sincerity and productivity are not only essential for big organizations, but also it is applicable to small ones as well as to any individuals. Every organization has the basic need to manage time and attendance of employees to increase work efficiency through punctuality and to eliminate the possibility of time fraud and buddy punching. Small organizations have misconception that it is wasteful to invest money in the biometric system. According to various case studies, negligible investment on biometric time and attendance management system not only increases productivity but also saves huge amount of money per year. Biometric system provides the most up-to-date and accurate data to payroll processing software which maintain process and prepare payroll. It reduces the payroll errors.

Biometrics Is a Complicated System
The basic purpose of biometric system is to simplify life, not make it more complicated. Biometric systems are used very easily and conveniently than other manual time and attendance systems. As administrative monitoring and supervision are required in case of traditional time and attendance system, there is no need of administrative burden. It has an efficient interface which can be managed and monitored remotely. Workers feel their accountability autonomously, even though the administrator or manager is not present. Accountability toward work and responsibility can directly increase productivity as well as efficiency of companies.

7.3 Vulnerability or Susceptibility Points of a Biometric System

Biometrics accompanies a lot of various mind-boggling concerns which need cautious thought. The above mentioned concerns delineate various potential entanglements and ideally give a sign of the outcomes of badly thought about biometrics arrangements.

Researchers and analysts recognized eight places in the nonexclusive biometric framework where assaults may happen [6, 17].

An important factor to utilize biometrics is to validate the genuineness of an individual. The weaknesses of a biometric framework are as follows:

1. Fake biometrics – It is represented at sensor. Reproduction of biometric traits is as an input to the system. Artificial fake finger, face mask, duplicate copy of signature, etc. are the examples of this mode of attack. Biometric technology that uses liveness detection technology is always difficult to spoof. This is not actually

the biometrics rather the technology that is being used behind the process. There are two types of biometric technology. One type stores images and identifies individuals matching those stored images. On the other hand, other types (mostly the sophisticated ones) read the biometric data and convert them into binary code. This sophisticated system applies liveness detection technology to check the liveness of the object. As a result, it becomes nearly impossible to fool the system. For example, if our biometric technology uses image processing system, then anyone can fool it just by holding the picture in front of the identification machine, like pictures of the vein, a picture of our face, etc. On the other hand, if that technology uses liveness detection technology, then the system will try to find out the liveness of the object, so in that case the picture will be detected as a nonliving thing, and the system will reject the entry. So, in my point of view, it is the technology underneath that ensures the accuracy of the system, not the biometrics. However, using multimodal biometric authentication system can also help to skip those spoofing too. One can use silicon to fake the fingerprint device but hardly can use the iris. So if both of the systems are used to double-check, that can surely make the process strong and tough to fool the system. The iris and retina biometric systems are more difficult to spoof [12].

2. Replay old data – In this mode of attack, already-saved digitized signal of biometrics is reprocessed to the biometric framework bypassing the sensor. In case of fingerprint recognition system or speech recognition systems, previously stored image of fingerprints and recorded audio signal are replayed, respectively.

3. Override feature extractor – In this type of attack, a malicious code called Trojan horse may alter the features of biometrics. Instead of the actual stored value gathered from the sensor, the feature extractor generates manipulated data which is chosen by the intruder.

4. Manipulation with captured features of biometrics – In this, actual captured feature set may be altered with fake feature set. Regularly the two phases of highlight extraction and matcher are indistinguishable, and this method of assault is incredibly troublesome. In any case, if particulars are transmitted to a remote matcher (state, over the Internet), this danger is genuine. One could "snoop" on the TCP/IP (Transmission Control Protocol/Internet Protocol) stack and modify certain bundles.

5. Override matcher – It may be possible to corrupt the matcher module of biometrics by the attacker, so that the predetermined match stacks instead of the original set are produced.

6. Modify template – Changing in any event one design in the database could result in either affirming fake identity or declining authorized individual.

7. Intercept the channel – Information which is collected through pathway from database server of stored template and matcher may be altered by the intruder.

8. Override conclusion of system – A hacker can affect the final decision process of biometric system. Authentication function may be overridden by the intruder.

7.4   Matter of Concerns of Biometrics

As biometric frameworks provide more noteworthy dimensions of confidentiality, different assaults exist to increase unapproved access to a framework that is ensured by biometric verification. Sadly, biometric validation is definitely not an ideal framework. Various issues of the system are discussed here.

7.4.1   Biometric Framework Configuration Concerns

Biometrics which is perpetually connected with confidentiality and privacy ought to be sensibly protected and reliable. A portion of the biometric protection concerns are:

- Maverick transducers and unapproved securing of biometric tests
- Interchanges security among transducer, matchers, and the databases of templates
- Exactness
- Processing throughput
- Versatility
- Value of system
- Security

7.4.2   Confirmation

Rather than entering a secret word, biometric validation confirms your character by checking your special natural features. Various primary concerns to be addressed amid the assessment of biometrics for confirmation frameworks are discussed here. Basically we have to recognize if the framework is intended for distinguishing clients or validating clients. Recognizable proof of a client is a significantly more troublesome assignment.

Biometric verification confirms the query as to "who are you." It can be attached to both computerized and physical states. This arrangement is being used in various applications, viz., border security, identification of criminals in law enforcement, entitlement programs, etc. Interestingly, identification does not include a case of character by any stretch of the imagination. Rather, the framework is given a lot of (preferably complete) qualifications and solicited to analyze this set from certifications against the clients it is aware of, restoring an outcome, which recognizes the client being referred to.

This is a well-known one-to-many test, and it should be evident that this kind of test is both more work concentrated for the system and progressively subject to having a wide extent of credits to differentiate customers. Biometric distinguishing proof as a general rule compares to the present circumstance to a situation where an affiliation needs to perceive a person. The affiliation gets a biometrics from that individual and thereafter glances through a biometric storage facility attempting to precisely perceive the person. The biometric storage facility could be supervised by a law approval office, for instance, the Integrated Automated Fingerprint System (IAFIS) continuously run by the FBI in the USA or be a bit of a national character structure like India's UIDAI.

Biometric confirmation or authentication poses the inquiry "would you be able to demonstrate your identity" and is prevalently identified with verification of character in advanced situations. A framework will move somebody to demonstrate their character, and the individual needs to react so as to permit them access to a framework or administration. Confirmation is basically the check that recognized the authorized client. The client may be called confirmed if there is a match. It is called as a synchronized test.

7.4.3 Liveness Detection

With the expansion of web extortion and fraud every day, specialist organizations need an approach to guarantee that their administrations can't be undermined. Particularly in unsupervised confirmation circumstances, liveness identification is required to decide the client's quality. Biometric validation frameworks need to avoid advanced satirizing difficulties from replay assaults. Liveness testing is turning into an essential piece of biometric frameworks. Liveness identification has the expect to recognize a living and amid the biometric verification process present individual thusly and to repulse ridiculing assaults at the information catch subsystem. Techniques are anticipated to personifying biometric structures dynamically tedious. The technique that is discussed here is the assurance of liveness. To choose whether an individual is live when they present their biometric data to a structure can be a problematic task to automate in a way that is qualified to customers and conceivable to execute. Numerous techniques require one to be alive, for example, detection of temperature, beat oximetry, location of heartbeat in fingertip, electrocardiogram, reaction of dielectric, and impedance. Each technique possesses difficulties that make it hard to be computerized and incorporated into frameworks in the most straightforward manner conceivable. For instance, the additional hardware requisite to play out a portion of these tests, for example, electrocardiogram, can be costly and awkward for the client [12].

7.4.4 Collapse Rates

In case of day-to-day running and configuration of biometric system, collapse rate is critically considered. Two sorts of disappointment rates are ought to be taken care of: false acknowledgment rates and false dismissal rates. These parameters (false rate) are a module how absolutely the framework endeavors to check every client against the qualities enrolled for them. In this manner, a system that is designed and selected exceptionally exact and contains extremely low false acknowledgment rates will constantly be playing out a high false dismissal rates, in respect to a reasonable framework. Poorer administration of system cannot convey the positive client experience.

A trained security administrator tries to make an arrangement with low false acknowledgment rates and overlook the bogus dismissal rates. This is anything but a reasonable alternative in all cases, and the setup depends emphatically on the nature in which the biometrics-based framework is to be developed. In a military application, where security is an essential, the system provides high false rejection rate. In a business application, it is less degree for the affirmation of deferrals and difficulties related with false expulsion rates. Common factors of failures incorporate the accompanying:

- Inappropriate innovation decisions
- Lack of affectability to client observations and prerequisites
- Presumption of an issue which is nonexistence
- Insufficient encompassing
- Improper utilization of biometrics where different advancements would better take care of the issue
- Insufficient business case
- Inadvertent comprehension of population issues

7.4.5   Circumvention and Repudiation

Circumvention is a major concern that refers to how easily authentication system gets fooled. While biometric system is implemented, its weakest thread has to be considered, as just like a chain its weakest part determines the security level of the system. That means it is as secure as its weakest part. Intruders try to attack its weakest part not its strongest module. An interloper may deceitfully access the framework by going around the biometric matcher and examine delicate information, for example, medicinal records related to a truly enlisted client. Other than abusing the protection of the enlisted client, the impostor can change touchy information including biometric data.

Repudiation refers deniability. Intruders repudiate their involvements. Non-repudiation property is to be inculcated into biometric system to protect against their denial and solve any disputes among parties.

7.4.6 Handicapped Non-registrable Users

It may be possible that in the utilization of biometrics in a particular framework, few clients may be found who could not be enrolled in the framework due to their physical disability or handicappedness. In such cases an auxiliary arrangement should be there to take care of this. It is essential not to reject client with physical handicappedness.

The procedure of non-registrable user changes entrust on the biometric system which is used. Indeed, even without real physical damage, a few components can endure specifically areas of the populace.

7.4.7 Adaptability

Adaptability of biometric framework is the general issue in which it is important that arrangement provided by the manufacturer be squeezed to demonstrate that the arrangement offered will be suitably versatile.

In spite of that, the uses of biometric technology face a lot of ethical concerns, the primary issues being the singular security, the conflict with one's feelings and values, and the collection, affirmation, and usage of individual biometric data. The human rights type organizations contend that biometric mechanism impairs the human rights for assurance and anonymity. It causes disruption and has the capability to have authentic effect on entity flexibility and democracy privileges. Also, some extra crucial information about the individual may be provided by recognition system. For example, some physical traits like retinal pattern may provide some medical data which can be misused. The innovation may soon get obsolete as it can be hacked. However due to many concerns and threats in the world, viz., terrorism attack, theft of identity, fraud, illegal immigration, misrepresentation, and crime prevention and detection concerns, it is necessary to inculcate techniques to protect individual identification for further ID and checking. Meanwhile since 9/11 the biometric development has advanced hugely. The hardware has improved in plan and exactness, the expenses have dropped, and, in like manner, biometrics has cemented its place in the security world.

An inquiry which regularly gets posed is: "The thing that occurs if my Biometric Template gets stolen or hacked into. Will I become a casualty of Identity Theft?" From a general outlook, there isn't much which can truly occur, should this situation occur. For example, if you think logically, you observed that it is bit difficult to get meaningful information from the pattern of zeros and ones and/or probability curve. It isn't exactly equivalent to taking credit card number. Likewise, every vender of biometrics has their own exclusive, numerical enrollment and matching algorithm, so taking a layout and putting into another framework are just not possible. In any case, if one somehow managed to burrow further at the specialized dimension, biometric templates are much the same as whatever other innovation, which are inclined to disappointments, hacking, robbery, and at granular-could to a specific degree be figured out.

Biometric templates may be at high risk to theft and hacking under the below-mentioned four critical areas:

1. After the creation of templates which includes both checking and registration.
2. System in which database of biometric templates which depends on the used technology of biometrics is configured.
3. Biometric templates are transmitted from the biometric system to the database server in client-server model.
4. In cloud computing structure, biometric templates are stored into third party.

Without considering the social impacts over biometric, it is bit hard to the best execution of System. Because of growing security threats, it is not easy to accept the credibility of biometric framework. The expanding utilization of biometric frameworks has wide social consequences, and one all-encompassing thought is proportionality disenfranchisement covert (secrete).

Notwithstanding the way that biometric structures can be useful, the possibly enduring relationship of biometric attributes with an entity, their possible use for remote region, and their association with character records may raise social and lawful concerns. Such kind of challenges can impact a structure's affirmation with customers, their execution, or the choice on regardless of whether to utilize it. Biometric acknowledgment likewise causes huge legitimate concerns of repudiation, expert, unwavering quality, and, obviously, security. Eventually, social and legitimate elements are basic and ought to be considered in the plan, advancement, and sending of biometric acknowledgment frameworks.

7.5 Impediments of Biometrics

Verifying biometric data and guaranteeing the security of individual personalities are a developing worry in the present society. Conventional validation schemes for the most part use tokens or rely upon some secret information controlled by the user for checking his or her personality. These customary-based procedures have been extremely well known and have a few constraints. Conventional-based methodologies, for example, token, and information-based methodologies can't separate between an approved client and an individual approaching the tokens or passwords. Learning-based verification frameworks expect client to recollect and deal with various passwords/stick numbers, which results in client burden. The restrictions of customary validation strategies can without much of a stretch be overwhelmed by biometrics-based verification plans utilizing fingerprints, face acknowledgment, and so on while offering ease of use focal points, for example, client accommodation, as the client does not need to recollect various passwords and their related cards. All things considered, a great many people who have more than one bank card have stirred up their stick numbers. Notwithstanding, in spite of all the undeniable focal points, analysts at the Biometric Research Laboratory, BRL, inside Namibia Biometric Systems are quick to raise a few security and protection worries as illustrated beneath.

Biometrics is certainly not a mystery Unlike conventional traits, i.e., PIN and secret keys for ciphertext, that are just noted by the user, biometrics, for instance, face and fingerprints, can indeed be recorded and possibly mishandled by biometric experts without the customer's consent. Our analysts at BRL are quick to diagram that there have been a few examples where fake fingerprints have been utilized to evade biometric security frameworks. Some physical and behavioral traits, namely, face, voice, etc., are in like manner vulnerable against being gotten without the customer's express data [9].

Attacks to user interface and countermeasures Biometric system may be unfit if it is not able to distinguish fake and certifiable attribute. Intruders can tamper into the framework with bogus attributes like face mask and artificial fingerprint. Liveness discovery is a successful countermeasure to counterfeit biometric assaults.

Tracking Almost certainly, the equivalent biometrics might be used for various applications and fields, and the customer can possibly be pursued if affiliations interest and offer their specific biometric databases, while regular affirmation plans require the customer to keep up different characters to prevent following. The way that a biometrics proceeds as before presents a security concern. Biometric data may be secured on adaptable media, for instance, sharp cards if they will be used in affirmation mode. This ensures the data can't be used without the customer's own one-of-a-kind endorsement, despite what happens with data set away in a central database. Biometric confirmation/ID likewise can be acknowledged via remote access, by sending the biometric images and/or layout using a system to a gadget which will process further. It requires an exceptionally safe association. To guarantee that the transmitted information has not been undermined, watermarking can be utilized for this situation.

Biometrics can't be dropped Conventional security traits like passwords, PIN, etc. can be reset whenever needed. The same can't be changed about a person's biometric. Be that as it may, biometrics is an exclusively inherent character of the client and can't be removed whenever used.

Biometric recognizable proof can be a measurable procedure Varieties in conditions among enrolment and procurement just as substantial changes (transitory or perpetual) mentioned that 100% match is not possible. For a password, secret phrases, or a PIN, the appropriate response specified is accurately equivalent to the one that has been confine, or it isn't – the minimal variation is a purpose behind rejection; on the other hand, in the case of biometrics, there is no unambiguous line between a match and a non-match. Regardless of whether a match exists depends in this way not simply on the two informational indexes to be looked at, yet furthermore what safety buffer is considered average.

Obviously, smart cards may be misplaced or stolen. Thus, stored information of these must be encrypted and backed up. In any case, if the data is stolen, it is important to have the capacity to reject it and to deliver another template which could be utilized for further recognizable proof. Denial is simple when managing pin codes or passwords yet not with biometric attributes as we can't change our irises or our fingerprints.

Cancellable biometrics is another exploration field and some fundamental recommendations have been made. It is conceivable to create new facial pictures for an individual by sifting the first picture. Cancellable biometrics gives biometric frameworks, hypothetically, the capacity to re-issue biometric signals. Major advantage of cancellable biometrics is that various substances and various applications utilize distinctive changes for similar signs. This avoids the sharing between databases of various substances. For instance, a law authorization organization will utilize one change for a unique finger impression filter, and a business element will utilize an alternate change for a similar unique finger impression examine. This thought of "decent variety" makes cross-coordinating inconceivable. The dealer takes the biometric information from the client and analyzes it to a change from one of the change databases related with a specific administration. This should ease security worries as various changes are held in various databases per element [7, 8].

7.6 Challenges, Difficulties, and Issues of Biometric System

Biometric systems face two kinds of security threats. The first is the use of biometrics to guarantee and offer security to information systems. For what sorts of usages and in which space is a technique combining biometric propels commonly appropriate? Expecting that a biometric system is setup, another security threat is the protection, reliability, and steadfastness of the structure itself. Data confidentiality inquire about is required that tends to the novel concerns of biometric frameworks, viz avoiding irruption dependent on the presentation of fake biometrics, the replay of starting at now put away biometric characteristics tests. Structuring a protected verification framework can be a serious testing task. The issue of recognizing vulnerabilities and effectively executing countermeasures is a difficult one. It is obvious that a superior technique for investigating vulnerabilities during the plan procedure is required with the goal that framework architects will be more averse to neglect vulnerabilities [10, 11, 15].

7.6.1 Needs of Multi-Model Biometrics

Most biometric frameworks conveyed in genuine applications are unimodal, i.e., they depend on the proof of a solitary wellspring of data for confirmation (e.g., single unique finger impression or face). These frameworks require to fight with

Challenges and issues of various biometric system

Biometric identifier	Features/recognition approach	Challenge and issues	Pros	Cons
Fingerprint	Ridge patterns	Skin qualities (wet or dry, scratch, wounds, age, and release) influence the inspecting procedure Contacting the sensor causes distortion during the procurement procedure Some obstacles like social reputation, health issues, lack of awareness, etc [4].	Widespread uses in forensic sciences and law authorization situations It is the only authentic identification since B.C. according to archeological evidences It fits to examine criminological Low-cost procurement gadgets requiring little space Large inheritance database exists	Unique finger impression can be reproduced in latex utilizing an item moved by the individual Degradation of optical sensor due to contact of figure Quality relies upon age, occupation, and way of life of the person Technical issues because of the examining procedure Cannot be utilized for fingerless individuals
Face	Face localization procedure	Physical appearance (aging, expression, hairdo, glasses, makeup, disguise, and so forth) Picture conditions (brightening changes and camera qualities) Variety of facial expression may affect the performance	Easy use and non-intrusion sampling process Not requiring any consent from object Widely acceptable sensors Very easily identified individuals	Controlled lighting conditions required Obstructed by hair, glasses, caps, and so on Highly sensitive to the changes in light, posture, articulation, and maturing
Voice	Frequency-based features Voice of every person has distinct pitch, tone, and volume which are recognized	Background commotion Temporal change of voice because of Well-being conditions and speaker passionate state Unknown obtaining specification (channel, mouthpiece) A voice biometric trait can be corrupted by the physical and energetic state of the subject similarly as by regular issues	Easily accepted by user There is no physical contact There is no need of expensive sensors; just take simple telephone or microphone	Age, medical conditions, and emotional conditions affect the voice As voice biometric trait is behavioral, it is not stable Background noise and playback spoofing also affect voice quality

Iris	Random pattern of the iris	Iris is generally not clearly visible; it is covered by eyelashes, contact lenses, glasses, eyelids, and reflections from the cornea. Due to uncontrolled light, accuracy of scanner can be affected. Iris sensors are costlier than other biometrics. Difficult to recognize physically disabled or non-registrable users	Performance is quite impressive in terms of false accept rates. There is no harm or discomfort to person. Contactless inspecting procedure. Public acknowledgment. Iris is internal organ which is stable and doesn't vary with age. Low preparing costs	Few inheritance databases. User collaboration. Easily veiled by eyelashes, focal points, reflections, and so on. Cannot be utilized in legal applications like forensics. Cannot be examined for certain disabled persons (missing one or the two eyes)
Palm geometry	Wrinkles and texture between finger and wrist are recognized	Palm geometry varies throughout the lifetime. Variant to hand introduction (method for situating it on the sensor). Collection of record may be affected by uncontrolled environment with illumination and position of palm [1]	Good-quality images are adapted for criminal detection. Poor-resolution images are reasonable for commercial and public application like access control. Potentially powerful than fingerprint recognition. Widely acceptable and large database	Scanner size is bulky. Person's consent is required. Quality of sensors degrades due to direct contact to sensor. Not clearly distinguished individuals. Fake traits can be prepared
Hand geometry	Geometry structure of hand composed of finger's structure, palm size, length, etc.	Since it isn't extremely utilized for distinctive verification, yet rather in a check mode. It may not be the same in the midst of the improvement time of youths. Limitations in mastery (joint pain) or even ornaments may affect the correct data	Non-intrusive and contactless recognition. Affordable equipment in the commercial access control area. It is simple method to get high efficiency. Great user's acceptance	Hand geometry varies as children grow. Used for verification only not for identification. Lot of storage is required

(continued)

Biometric identifier	Features/recognition approach	Challenge and issues	Pros	Cons
Signature	Image-based as well as time notation features are included in which disconnected marks and number of vertical and inclined slopes are recognized through computer vision or neural network techniques	Due to dependency on the emotion and time, there are non-linear changes. Changes gradually over time. Signature verification is a bit difficult, whether it is genuine or forged	Frequently used authentic method of written documents. Most popular authorization technique known as seal of approval. Oldest means of verification	Dynamic signature needs costly equipment. Training required for utilizing electronic mark gadgets. Not stable biometrics particularly for the individuals who couldn't keep up steady composition
DNA	It is found inside the nucleus and follows the law of Mendelian inheritance	DNA recognition process is very complicated involving expert's skill. It is a very time-consuming process. Limited scope, viz., forensic applications. Highly intrusive method; person doesn't easily get convinced [19]	Due to unique property, it is highly reliable. It provides accurate recognition as compared to others. Sophisticated method for forensic applications. DNA doesn't change throughout the person's life	Difficult to distinguish identical twins. It is an expensive and time-taking process. Limited scopes. Person's consent is required. Highly experts or trained persons are needed
Retina recognition	Recognize blood vessel pattern which is in the back of the eye	It is a cumbersome process to capture images due to focus on a particular point in the visual part of eyepiece. It is dependent on the cooperation of the user; user must remain still in front of the embedded lens of the device	Most secure biometrics due to stable physical trait structure for lifetime; not possible to alter. Impossible to replace fake retina. Robust matching capacity	Require high-quality image so expensive sensors are needed. Not very user-friendly; user's support is needed. Users are not comfortable with eye-related technology; they fear about damage of the eye. Limited uses and scopes

Thermograms	Anticipation of the thermal energy which is emitted from the human body	It cannot be forged or altered It is not affected by shadowing or illumination circumstances Thermograms are invariant to disguise Thermograms cannot be forged or modified No physical contact is required It is a robust recognition process It is not affected by aging factors Widely used to diagnose vascular disease, breast tumors, or various medical diseases	It really needs expensive sensors Still in the developing stage Limited scopes mostly in medical community Very difficult to collect data due to unsuitability to acquire image at low temperature
	Specialized sensor cameras are required		
Gait	Identifying people by their pattern of walk; recognize articulate joints	Due to behavioral traits, gait may not stay invariant over a long span of time The principal challenge concern in such biometrics is acquired from individuals tracking issue, where enlightenment variation, shadowing, impediments, and so forth truly sway the execution of walk acknowledgment frameworks Due to body weight, major injury, change in clothing, and illumination, pose decreases performance	It is widely acceptable because of contactless It is ideal for low security applications Its pertinence to acknowledgment of individuals at a separation in video pictures, with the goal that secret ID may be plausible
			Lack of complete accuracy Not invariant with time Computationally expensive since it requires ore computations It is not very distinctive

(continued)

Biometric identifier	Features/recognition approach	Challenge and issues	Pros	Cons
Keystroke	Recognizable proof of an individual by his own composing style	It is as similar as speaker acknowledgment which is based on fixed content, for example requesting that the client type certain content and concentrate highlights from his composing style Keystrokes pattern may vary due to illness, lack of skill, tiredness, etc.	Needs no special hardware aids or expensive sensors Cheapest mode to recognize No training is required for enrolling or registering their live samples	Still require to develop robustness and distinctive Limited use due to lack of computer literacy
Ear recognition	Structure of cartilaginous of the pinna is recognized	It can now and again be covered up with hair, top, turban, suppressor, scarf, and hoops It is not widely acceptable and believed recognition	Shape and appearance are fixed throughout individual's lifetime Shape is not changed with facial expressions It is the most stable and authenticates biometric system There is very less computational complexity Faster identifications There is no need of physical contact	Uncontrolled illumination and variant pose affect the performance of the ear recognition Wearing of hat, earring, and scarf also affects performance significantly
Skin reflectance	Reflectance spectrum of skin	High-resolution, high-accuracy spectrograph under controlled lighting condition is required Penetration of light depends upon the wavelength of light	Cannot be forged Highly reliable since no two people have the same reflectance spectrum of skin	Highly person dependent Not very user-friendly Expensive technology

Lip motion recognition	Biometric traits which capture distinct characteristic of people like the size of upper and lower lips, furrows, grooves, etc. as well the motion of lips It is a physical as well as behavioral biometric trait	Lips are highly deformable so robustness and accuracy are very difficult to achieve	It is less susceptible to background noise Very useful for deaf and dumb persons It is language independent Under controlled lighting condition, it can be fairly well extracted Distinctive and unchangeable attributes for every examined person Template size is small Interaction of user is not necessary It can be hybrid: Lips-voice or lips-face	Needs more attention for hybrid system Variations (smile) may cause difficulty in recognition
Body odor	Due to chemical composition, each object spreads body odor Based on substance identification	Require calibration to senses Artificial nose is not yet sufficiently advanced to do all the activity There are no accessible business applications available yet Difficult faculties to measure Deodorants and aromas could bring down the uniqueness	Identification is possible by mixture of odors by recognizing the mixture's components No need to take consent from persons Non-intrusive Cheapest way to test quality in food industries It is treated as a diagnostic tool in medical industries	Phony noses are not comfortable to do the entire operation Quantification interpretation is troublesome Distinctiveness is reduced by deodorants and perfumes
Writer recognition	It implements right division into characters and finds the most possible words	Require earlier limitation and division of the applicable data, which is normally performed intelligently by a human client An insignificant measure of penmanship (e.g., a passage containing a couple of content lines) is important so as to determine stable highlights unfeeling toward the content substance of the samples	It cannot be imitated It is easily used in forensic and historic document analysis Have no privacy rights issues	The same individual can have inconsistent writing Has very limited market

a variety of challenges, viz., noise in detected data, high intra-class variation and low between class varieties, non-all inclusiveness, parody assaults, limited degrees of opportunity, and unsatisfactory blunder rates. Due to these serious constraints, the unimodal biometric systems when need to be deployed in nations like India, huge population size demands for much more competent field solutions (especially toward ensuring 100% population coverage). A few disadvantages of unimodal biometric frameworks can be reasonably settled by multi-biometric framework in which numerous wellsprings of data are inundated. Such biometric frameworks are progressively solid because of capturing and investigating various biometric attributes which are autonomous bits of proof.

Various criteria, namely, individuality, repeatability, user-friendliness, and acceptance, are hardly satisfied in any one of the biometric systems completely. Some of the limitations of biometrics include similarities among different people; change in attributes over time, i.e., aging and physical limitations; acceptability by the user; and many people not having all the characteristics. Durability is the major issue in biometric system. The most recent research demonstrates utilizing a mix of biometric roads for human acknowledgment is unmistakably more successful than the unimodal biometric frameworks. Thus, a multi-biometric engine is needed for real-time biometric recognition systems, which can effectively alleviate the problems observed in unimodal biometric systems. The key field issues of unimodal systems are non-universality (which will lead to insufficient population coverage, due to high failure-to-enroll rate) and spoofing.

Multimodal biometric systems utilize one or more physical more than one physiological or behavioral trait for registration, confirmation, or ID. The basic purpose of combining various processes is to improve acceptance rate.

The aim of multi-biometrics is to reduce the false rejection rate (FRR), failure-to-enroll rate, and sensitiveness of device.

The need of multimodal systems can be very well understood by studying the following key limitations of the unimodal systems.

Unimodal biometric systems need to fight with an arrangement of issues, for instance, boisterous data, intra-class assortments and restricted degrees of chance, non-comprehensiveness, spoof attack, and unacceptable mistake rates. A portion of mentioned constraints can be influenced by conveying multimodal biometric frameworks that coordinate the evidence introduced by different wellsprings of data [5]. Here in the following sections, different situations are discussed which are possible in multimodal biometric framework. The element of combination and coordination system must be appropriately received to merge data.

1. *Noise in sensed data*: Due to flawed obtaining conditions, the caught biometric characteristics may be loud or contorted. Such varieties in biometric data may create false dismissals/acknowledgments in the database, i.e., an enlisted client of the framework may be inaccurately dismissed, or something else, an impostor may be erroneously acknowledged. In face advancements, enlightenment conditions may influence the nature of the caught face pictures, or too brilliant

encompassing light may influence the execution of the unique mark optical sensor.

2. *Non-all inclusiveness:* Although biometric attributes are relied upon to exist among each single individual of a given populace, there are a couple of unique cases, where an individual can't give a specific biometrics. For example, due to pathological conditions of the eye, iris images are probably not be procured, or in working environments characterized by manual activities, the fingerprint's structure might even be disappeared.

3. *Upper bound on identification accuracy:* The accuracy in biometric systems might be improved by developing more robust techniques, though there exists an upper bound on the system's precision. It entrusts basically in the magnitude and quality of attributive patterns which can be modeled using a template. The discriminative capability of a template is constrained by two factors; they are, namely, the intra- and inter-class disparity in an individual. While the former refers to the variations among feature sets of a same individual, the latter refers to the variations among feature sets belonging to different individuals. In some cases, there exists a high intra-class variation, which means that biometric features belonging to the same individual vary substantially, whereas low inter-class disparities mean that biometric features of diverse persons may emerge quite similar [13]. The accuracy of biometric systems may be affected by high intra-class and low inter-class disparities.

4. *Spoof attacks:* Biometric frameworks are defenseless against parody assaults; hence, by consolidating various sorts of biometric attributes into a solitary application, the achievement of farce assaults will be undermined [14].

Why Is Multiple Traits Combined?

Each recently proposed physical biometric identifier has ended up having issues of some sort – regardless of whether it be unique mark, iris, facial, voice, hand geometry, ear shape, retina, or the rest – everyone presents us with a trade-off sooner or later.

Be that as it may, we should ask: What qualities would the perfect biometrics have? Obviously, to a degree that relies upon your application; however there are some key properties that we can accept that are attractive for all physical biometric identifiers. These may be extensively gathered into the regions of security and reasonableness, as clarified beneath:

1. *Security*

Security is the major concerned to use biometrics. There is no use of biometrics without it.

There are two fundamental parts of security that our optimal biometrics must fulfill:

Resistance to Fake

An "internal" biometrics will customarily be more secure than an "outside" biometric, since it will fundamentally be hard to duplicate or change.

A case of the potential instability of outer biometrics was found by Japanese scientists who had the capacity to lift fingerprints from glass and verify to unique mark scanners utilizing counterfeit fingers produced using family fixings.

Accuracy

A low false acknowledgment rate (FAR) is vital to security; it is important that unapproved people are not misidentified as approved. The FAR relies upon a few components, for example, the peculiarity of the picked biometrics between people (basically, its uniqueness), the capacity to catch the biometric data precisely, and the capacity to coordinate it accurately. Starting here of view, the iris is a decent biometrics; there are a high level of irregularity and multifaceted nature in iris designs that supports its uniqueness and peculiarity between people.

Be that as it may, the FAR spreads just a single side of exactness. The bogus dismissal rate (FRR – erroneously dismissing people who are really approved) likewise matters, yet is even more a common sense issue, as examined in the following sub-area.

2. *Practicality*

Clearly, we can propose a validation procedure that is as secure as we can imagine, yet except if it is additionally functional, it will never succeed. Common sense and security are frequently inconsistent with one another, and any last framework will be an exchange off between them. The time when we are set up to make the trade-off will rely upon the estimation of what we are ensuring – the security frameworks guarding the Crown Jewels would not be handy to execute in the normal UK home. In a perfect world, our picked biometrics ought to have attributes that limit the requirement for trade-off by improving both security and common sense.

The following practicality factors have to be considered:

- *Speed.* Biometric system ought to be quick to use and processed practically. Speed factor plays a significant role. A representation of this is given by a UK school that actualized iris acknowledgment for their midday food payment system. Following a year the system was relinquished for being excessively slow. "We don't need understudies' dinners getting cold while they hold up in the line," said the head teacher.
- *Accuracy.* A framework that rejects authentic clients might be secure, yet is unquestionably badly designed. The ideal biometrics will never dismiss an approved individual (zero FRR) and never acknowledge an unapproved singular (zero FAR). Among innovations with the most noticeably terrible FRRs are faces and voice acknowledgment.

Some of the time unessential variables can influence the exactness of a biometric strategy. A voice recognition framework may disregard to perceive a customer with cold. A distinctive spot framework can be sensitive to dirt or grease or scraped area on the finger. An iris framework might be influenced by the nearness or nonattendance of exhibitions. A perfect biometrics will be as harsh as conceivable to superfluous components.

- *Cost*. There will dependably be a trade-off between the security a framework offers and its expense. For instance, fingerprint peruses are currently aware, ease item; in any case, they don't offer an abnormal state of security. A perfect biometrics will be "financially savvy," ready to offer a moderately abnormal state of security at a generally minimal effort.
- *Size*. Numerous security applications have size limitations, for example, PC login or entryway get to. The need to accommodate a substantial reader gadget (e.g., on account of palm acknowledgment) is a viable issue.
- *Convenience*. Obviously our optimal biometrics ought to be simple and helpful to utilize. People encounter this issue during utilization of iris recognition. The innovation utilized requests that individuals expel their glasses; shockingly for the shallow, this makes the "objective zone" difficult to see. Several minutes spent attempting and neglecting to get adjusted outcomes just in dissatisfaction
- *Enrolment*. Some biometrics are less reasonable than others: fingerprints can wear or wrinkle with age to the point where they become unusable; hanging eyelids can be an issue for iris acknowledgment. This is especially an issue for the debilitated, where the selected biometrics may not be available, or real position might be an issue. For example, a UK Passport Service consider demonstrated a 39% failure to enlist (FTE) for handicapped individuals utilizing iris acknowledgment.
- User *acknowledgment*. There are various reasons clients may oppose a biometric system:
    - *Privacy* concerns. For instance, it stresses that it may prompt remote tracking. A biometrics that can't be perused from a separation (in contrast to face, voice, iris) is ideal.
    - *Hygiene* issues – applies to contact systems, for example, fingerprint.

3. *A Trading-Off Position*

The majority of the standard biometrics has issues in at least one of the regions recorded previously. A biometrics that sparkles in a single zone will be a letdown in another. This guarantees that the decision of biometrics for a specific application will be a trade-off, ordinarily, a trade-off among security and common sense, and a wrong trade-off will unavoidably prompt the disappointment of a biometric venture.

7.7 Conclusion

This chapter deals with major challenges, issues, and social impacts of various techniques involved in identification. The biometric recognition systems are the automatic recognition system to overcome the drawbacks of traditional system. But it also has some limitations and can be overwhelmed with the development of biometric innovation.

Biometrics is a capable and empowering field, where different request meets and allow to continuously protect and careful world. There are different predominant biometric segments as of now, some with strong stories and some generally new frameworks. Each device has its very own characteristics and deficiencies. Right when suitably associated, biometrics can be used to experience coercion and certification that participation frameworks are direct and definite.

Utilizing a particular feature of biometrics may prompt great outcomes, yet there is no solid method to check the grouping. To accomplish hearty recognizable proof and confirmation, two diverse biometric highlights can be joined. A multimodal biometrics can give a progressively adjusted answer for the security and accommodation prerequisites of numerous applications.

Biometric advancements which are in progress have achieved about extended precision at a decreased cost; advances in biometrics are situating themselves as the establishment for some exceedingly secure recognizable proof and individual check arrangements.

Notwithstanding the huge advancement made in the course of recent years, biometric frameworks still need to figure with various issues, which outline the significance of growing new biometric preparing calculations just as the thought of novel information procurement systems. Without a doubt, the synchronous utilization of several biometrics would improve the exactness of an ID framework, viz., the use of palm prints can support the execution of hand geometry frameworks. In this manner, the advancement of biometric combination plans is a significant region of learning. The likelihood of utilizing biometric data to produce cryptographic keys is similar as developing territory of study. Consequently, there is a clear requirement for cutting-edge flag preparing, PC vision, and example acknowledgment procedures to convey the current biometric frameworks to development and take into consideration their substantial scale sending.

References

1. Z.M. Noh, A.R. Ramli, M. Iqbal Saripan, M. Hanafi, Overview and challenges of palm vein biometric system. Int. J. Biom. **8**(1) (2016), https://doi.org/10.1504/IJBM.2016.077102.
2. S. Asha, C. Chellappan, Biometrics: an overview of the technology, issues and applications. Int. J. Comput. Appl. **39**(10), 35–52 (2012)
3. R. Bolle, J.H. Connell, S. Pankanti, N.K. Ratha, A.W. Senior, *Guide to biometrics* (Springer, New York, 2004)
4. A.K. Jain, J. Feng, Latent finger print matching. IEEE Trans. Patt. Anal. Mac. Intel. **33**(1), 88–100 (2010)
5. A.K. Jain, A. Ross, S. Prabhakar, An introduction to biometric recognition. IEEE Trans. Circ. Syst. Video. Technol. **14**(1), 4–20 (2004)
6. N.K. Ratha, J.H. Connell, R.M. Bolle, Enhancing security and privacy in biometrics-based authentication systems. IBM Syst. J. **40**(3), 614–634 (2001)
7. B. Schneier, The uses and abuses of biometrics. Commun. ACM **42**(8), 136 (1999)
8. N.K. Ratha, R.M. Bolle, Smart card based authentication, in *Biometrics: Personal Identification in Networked Society*, ed. by A. K. Jain, R. M. Bolle, S. Pankanti, (Kluwer Academic Press, Boston, 1999), pp. 369–384

9. B. Schneier, in *Security Pitfalls in Cryptography*, Proceedings of the CardTech/SecureTech Conference, CardTech/SecureTech, (Bethesda, 1998), pp. 621–626

10. T. Sabhanayagam, V.P. Venkatesan, K. Senthamaraikannan, A comprehensive survey on various biometric systems, Int. J. Appl. Eng. Res **13**(5), 2276–2297 (2018) ISSN 0973-4562.

11. U. Uludag, S. Pankanti, S. Prabhakar, A. Jain, Biometric cryptosystems: issues and challenges. Proc. IEEE. **92**, 948–960 (2004)

12. P. Campisi, Security and privacy in biometrics: towards a holistic approach, in *Security and privacy in biometrics*, (Springer London, 2013), pp. 1–23

13. A. Adler, R. Carppelli, Template Security in: *Encyclopedia of Biometrics*, (Springer, 2009), pp. 1322–1327

14. D.C.L. Ngo, A.B.J. Teoh, J. Hu, *Biometric security* (Cambridge Scholars Publishing, England, 2015)

15. K. Jain, K. Nandakumar, A. Ross, 50 years of biometric research: accomplishments, challenges, and opportunities. Pattern. Recog. Let. **79**, 80–105 (2016)

16. K. Nandakumar, A. Jain, Biometric template protection: bridging the performance gap between theory and practice. IEEE Sig. Proc. Mag. **32**, 88–100 (2015)

17. N. Bartlow, B.Cukic, The vulnerabilities of biometric systems – an integrated look at old and new ideas. Technical Report (West Virginia University, 2005)

18. K. Jain, K. Nandakumar, A. Nagar, Biometric template security, EURASIP J. Adv. Sig. Proc. **2008**, 1–17, 579416 (2008), https://doi.org/10.1155/2008/579416.

19. S. Tiwari, J.N. Chourasia, V.S. Chourasia, A review of advancements in biometric systems, Int. J. Innov. Res. Adv. Eng. (IJRAE). **2**(1) 2015, ISSN 2349–2163

Chapter 8
Segmentation and Classification of Retina Images Using Wavelet Transform and Distance Measures

Ambaji S. Jadhav and Pushpa B. Patil

8.1 Introduction

It is observed that the diabetes disease covering over a large scale of population is creating an eye problem popularly known as "diabetic retinopathy (DR)." It is becoming now a common disease in a large scale of population.

Diabetes is spreading in the population at a very fast rate. Therefore, disease detection at early stage can avoid blindness in the patient. However, it requires a large number of specialists and infrastructure. To reduce the load of experts (doctors), there is a need of computerized automated technique so that disease detection should take less time. Technical screening system using computers is becoming a more flexible method and acting like a primary test which can quickly assist the doctors.

Rise in glucose levels related with diabetes is the known reason for diabetic retinopathy (DR). The DR is dynamic progressive sickness of the retinal layer that do not show any signs in the beginning of DR and over the time the effect of diabetic cumulative on the health and starts effects on the retina where disease progress is indicated by different lesions in retina. The diabetes disease does not create any problems to the patient when it starts, so there is a very rare chance that the patient undergoes diabetes test unless he or she has some other health-related problems. In the case of other health problems, medical experts doubt about the sugar level in the blood, and they may advice for blood test; however, by this time, diabetes might have progressed. Diabetic retinopathy may be classified briefly into three stages, namely:

The early stage where the disease has not progressed; it is also called nonproliferative diabetic retinopathy and is represented as NPDR.

A. S. Jadhav (✉) · P. B. Patil
B.L.D.E.A's V.P. Dr. P.G.H College of Engg. & Tech., Vijayapura, India

© Springer Nature Switzerland AG 2019
G.R. Sinha (ed.), *Advances in Biometrics*,
https://doi.org/10.1007/978-3-030-30436-2_8

The second stage where disease has made progression and effects on different organs is occurred this is also called as proliferative diabetic retinopathy and represented as (PDR) creates from impeded vessels that directs to retina layer thickening and arrangement of extra auxiliary vessels on the retina which may be close to the optic disc or over retina in different places. Diabetic retinopathy is described by various highlights that are conspicuous by the prepared spectator. The highlights of DR, for example, the quantity of small-scale aneurysms and spot hemorrhages, have been exhibited to associate with ailment seriousness and likely progress of the diabetes, at any rate for its beginning times. Such sores have a sensibly all around characterized by appearance of signs of diabetes on the layer of retina and helps in detection and the recognition of them gives valuable data.

The third stage is a very crucial stage where the disease has made much progress because of cumulative effect of diabetes; this is also called sight-threatening diabetic retinopathy represented as STDR.

It is additionally critical that DR is a controllable disorder all through illness movement initiating from the beginning stage. Whenever recognized near the beginning and at same properly treated then there is possibility of minimizing saving expenses and either decrease or stop the further progress of disease this avoids the patient from losing his eyesight. The detection and treatment of DR are tremendous asset depleters on governments and well-being frameworks around the world. Since the disease of diabetes controllable and treatable, diagnosing the disorder through fundus camera images is useful and progressively productive recognition and checking minimizes cost and money, doubtlessly computerized automated detection of diabetic retinopathy is a needed for a developing need. Figure 8.1 demonstrates the disordered labeled fundus image.

Many algorithms have been developed for computer-assisted detection of DR features like exudates, hemorrhages (HA), and microaneurysms (MA). Retinal blood vessels are the network of vasculature of the retina. It consists of twigs and

Fig. 8.1 Color fundus image with anatomical structures and disease annotated

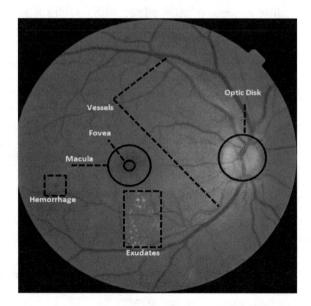

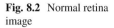

Fig. 8.2 Normal retina image

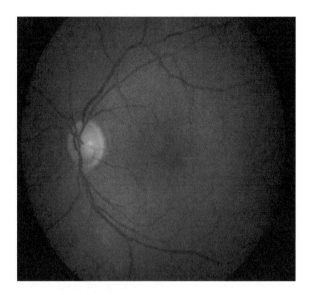

tributaries of the fundamental retinal channel and veins. The retinal arteries supply the oxygen-rich blood to the retinal layer of the eye. When any obstacle occurs in the retinal major artery (small twigs), the retina cells slowly start suffocating from oxygen deficiency which causes the DR.

Segmentation of blood veins of retinal images helps in the early diagnosis of the progress of diseases such as glaucoma, DR defect detection, defect detection, hemorrhage revealing, etc.

The configuration of the blood vessel is an observable feature in the retinal structure; it reflects measurable anomalies across the blood vein structure. Figure 8.2 shows a normal person's retina image that does not have any sign of diabetes.

The effect of hypertension is recognized from the blood circulatory framework. The effect of extra pressure or hypertension will develop diabetes, and in turn diabetes can affect other organs including retina part of the eye.

8.2 Structure of the Eye

The structure of the human eye is shown in Fig. 8.3, which consists of the retina as one of the layers which is responsible for visualization.

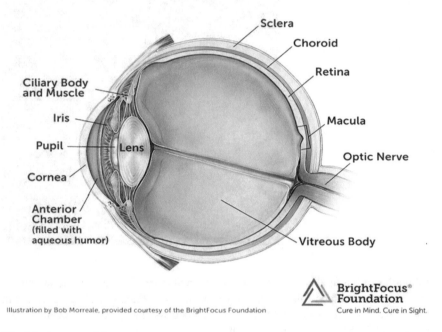

Fig. 8.3 Anatomy of the eye

8.2.1 Lexicon of Terms for the Eye

Front compartment This is also known as anterior chamber. This is a section of the eye connecting the cornea and the central point that consists of aqueous hilarity.

Aqueous hilarity This part of the eye is also called *aqueous humor. This contains* liquid produced in the eye.

Bruch's film This is situated in the retina linking the choroid layer which is a tough tissue that provides security to the eye and the retinal pigmented epithelium (RPE) layer, provides bear to the retina, and works as the cellar film of the RPE layer.

Ciliary body This is another important section of the eye, above the central point, that produces the aqueous humor.

Choroid This is a second layer or film coating that provides safety to the eye from external sources that can enter into it. It is at the back of the retina and contains veins that will nurture the retinal layer.

Cones This is another constituent of the eye which is responsible for differentiating fine details of the object to be viewed.

Cornea This is an exterior and simple structure located at the border of the eye that covers the iris, beginner, and front cavity. This is the eye's necessary light-centering arrangement.

Drusen This accumulates the yellowish supplementary cell waste items that collect inside and below the retinal layer that contains the pigmented epithelium.

Fovea This is the cavity down at the central point of the macula which gives the finest visual irregularity or sharpness.

Iris This is the somewhat ring type of tissue at the back of the cornea with the aim of directing the compute of light incoming to the eye by modifying the degree of the replacement.

Focal point This is the basic and simple structure balanced following the iris with the purpose of centering light in the region of the retina; it mainly gives a correction change in agreement with the vital centering arrangement of the eye.

Macula This is the sector of the eye on the crucial position of the retina so as to form spiky, clear directly forward vision.

Optic nerve This is the mass of nerve structured filaments on the backside of the eye so as to communicate visual messages beginning in the retina near the cerebrum.

Photoreceptors This is the illumination identifying nerve-like structured cells positioned in the retina.

Understudy The flexible opening at the focal point of the iris through which light enters the eye.

Retina This is the light insubstantially sensitive coating or layer of tissue so as to line up the backside of the eye.

Retinal pigmented epithelium (RPE) This is a film of cells with the intention to ensure and tolerate the retina. It also evacuates waste substance, forestalls new recruits vessel growth interested in the retinal coating, and ingests illumination not inspired by the light-sensitive cells. The actions turn away the capturing of the light in the eye and improve clarity of vision. The dimension of light entering into the eye is controlled by this layer.

Sclera This is the farthest outside coat of the eye which ensures the protection to the entire eyeball.

Trabecular meshwork This is soft tissue structure positioned close to the cornea from side to side which dilute amusingness flow out of the eye.

Vitreous This layer removes squeeze-like material to fill the eye starting from the crucial point to the backside of the eye.

8.2.2 Facts of Diabetic Eye Disease

1. Diabetic patient eye disease contains a meeting of eye circumstances that control persons with diabetes disease. The situations include mainly diabetic retinopathy (DR).
2. Everyone who is suffering from diabetes has diabetic eye disease, and it can lead to vision problem at any time of a person's life.
3. The eye disease caused by diabetes includes modification to retinal vessels which carry blood to different parts of the retina, thereby causing release of liquid-like protein.
4. In the total cases of visual problems, most are because of diabetic retinopathy.
5. DME is a main consequence of diabetes so as to create enlargement in the area of the retina called macula.
6. The way to controlling diabetes is maintaining a healthy diet and doing physical exercise regularly along with doctor-advised medicines. The effects of diabetes can slow downed and vision pran be postponed.
7. As the diabetes disease usually goes unobserved until some kind of health problem occurs, vision trouble can happen at any time, so persons with diabetes are supposed to get a complete eye check once in every year.
8. Near the beginning detection, appropriate treatment and appropriate follow-up deliberation of diabetic eye disease can make sure the prevention of vision loss.
9. Diabetic retinopathy disease may be treated by means of a small number of therapies, used single-handedly or in combination.
10. NEI supports do research to find out new methods for treatments for diabetic retinopathy and to give the impression of being at the viability of accessible treatments for a variety of patient gatherings.

8.3 Symptoms and Detection

An analogous scene or problem occurs as seen by the human being typical vision and with progressed diabetic retinopathy. The migrating red spots are created popularly known as hemorrhages that need concise treatment from medical experts.

The commencement time of diabetic retinopathy advances regularly without creating considerable side effects. The diabetes disease regularly advances without being noticed by the person till it creates a problem related to vision. The material kike fat and protein leaking from extraordinary retinal veins will originate the occurrence of floating spots. These spots will appear once in a while clearly.

Diabetic retinopathy and DME are determined throughout a comprehensive dilated eye examination that contains the following.

1. *Visual acuity testing*. This is one of the eye tests also called eye graph analysis which determines a person's capacity to notice at dissimilar separations.

2. *Tonometry.* This is another eye examination which finds or estimates weight within the eye.
3. *Pupil dilation.* This is one type of test where drops on the eye's exterior enlarge the person's eye, enabling a medical expert or doctor to check the retina along with optic nerve functioning.
4. *Optical coherence tomography (OCT).* This is one more type of eye test. This is similar to ultrasound process that utilizes light waves relatively than sound waves for holding images of tissues surrounding the eye part. OCT test provides exhaustive images related to tissues so as to be penetrated by light, like the eye.

A complete dilated eye assessment allows the medical doctor to verify the retina for:

1. Modification or new origin of blood vessels.
2. Blood vessels become weak and start leaking blood because of fat deposited in veins.
3. Enlargement happens in the macula (DME).
4. The lens of the eye gets changed.
5. Optic nerve becomes weak and injured.

On the off chance that extreme case of diabetes is assumed, a fluorescein in angiogram may be utilized to search for harmed or cracked veins. In this process of examination, a bright color is infused inside the circulation system, frequently into the section of the vein. Images of the retina are taken when color changes.

IOP ought to be examined particularly as soon as NVI is observed. Enlarged fundus image inspection ought to incorporate a macular assessment (examination with respect to contact focal point otherwise without contact focal point) for the purpose of investigating microaneurysms, discharge, bright lesions or white patches called soft or hard exudates, fiber or cotton wool areas, and retinal swelling also called macular edema. The point where blood vessels originate is popularly known as optic disc, and territory encompassing optic disc (one circle distance across) is to be inspected for nearness of strange fresh small vessel branches (newly created vessels around optic disc circle, NVD). Further the rest of the part of the retina ought to likewise be analyzed to observe nearness of irregular fresh vessel branches.

Macular Edema

Macular edema is nothing but swelling or thickening brought in the retina because of leaking of internal retina liquids. It is accepted to be an effect of high pressure created in retinal veins. DME or macular edema may be available with any dimension.

ETDRS Criteria for Clinically Significant Macular Edema (CSME)

• Retinal thickening at the center of the macula
• Retinal thickening and/or adjacent hard exudates at or within 500 μ of the center of the macula

8.3.1 Retinal Analysis

The retina is the deepest, light fragile layer or cover which is acting like a display device in eye structure of all living beings. The optic nerve structure of retina makes an arrangement for creating two dimensional picture of any object which is be viewed and interprets the object along with the creation of electrical neural signals and motivation to the cerebrum for proper recognition of the object which focused on retina. In terms of overall description, the retina serves as CCD in digital camera.

The retinal structure of human eye consists of a number of neurons which are interconnected for the purpose of conveying information from one layer to another; they are also called neurotransmitters. The neuron structure of the retina contains three different layers of neuron or neural cells, namely, the first one is image receptor neuron or cell, the second is bipolar neuron or cell, and the third is ganglion cell in the interior of the retina, which is entirely containing ten meticulous sections or layers, which include an exterior layer of pigmented epithelial neurons or cells. The foremost neural cells which are directly concerned with light entering in the eye are photoreceptor cells, and they are of two major types, popularly known as rods and cones. The main goal of rods is to reduce the excess light intensity and provide good contrast in terms of vision of object. The main responsibility of cones is to analyze and view shading.

Light falling on the retina starts with a course of synthetic and electrical signaling occasions those at last activate nerve driving forces which will be sent to different image focuses of the mind from beginning to end the filaments of the optic nerve. Neuron signals received from retina structure rods and cones will be processed by other neurons in the eye structure; these neuron's output will be the function of potential in the retina cells called ganglion cell axons which make up the optic nerve.

In any animals the progression of retina structure and the optic nerve start growing outside till it approaches the cerebrum, one of the main parts of the mind which is also called the brain. So keeping all those aspects into consideration, the retina can be projected as the main feature of the central sensory system (CNS). Therefore the retina is actually mind tissue.

8.3.2 Causes of an Occlusion

- Basically all the tissues in the retina require proteins and other nutrients for their proper functioning. The nutrients and proteins are supplied to retina cells by the blood itself, so blood flow mechanism is implemented by retinal veins which are also called blood vessels. As an effect of diabetes disease, the blood flowing through these blood vessels deposit some sticky-like material inside the blood vessels.

- The structure of blood vessels is such that they are very narrow and smooth so that whatever blood enters into them should flow very effectively and easily. In many of diabetic patients, the inner side of blood vessels is restricted because of the sticky material deposited; this will create some trouble for effective blood flow through these vessels.
- The patches or deposited materials inside the blood vessels are also called atherosclerotic plaques. Since these plaques are deposited frequently and solidifying later reduce size of vessels and in turn reduce the quantity of blood flow to retina cells, this leads to the pressure on the blood vessels.
- Blood vessels and corridors run along in all the areas of the retina firmly together in the back structure of retinal part of the eye. Because the inner wall of blood vessels gets sticky material after solidifying, it may reduce the size of blood vessels, thereby reducing the blood flow; sometimes if the deposited material is more, then there is a chance of causing blockage.

8.3.3   Risk Factors of Retinal Vessel Occlusion

Because the inner side of blood vessels gets narrowed or some time blocked, there are several risk factors which are to be considered. The following are some risk factors which arise because of occlusion.

- *Age factor:* In most of the cases of the retina, blood vessel occlusions will appear after the age of around 65 years.
- *Hypertension:* The flow of the blood to retina cells is with some normal pressure, and if the patient has high pressure, then there is a chance of occlusion happening.
- *High cholesterol:* Normally cholesterol has effect on many organs of the body; if the patient has a high level of cholesterol, then the blood flowing in the retina cells creates occlusion.
- *Blood clotting:* The probability of occurring occlusion in the blood vessels is high in people having blood clotting disease.
- *Glaucoma:* The result of occlusions can create increased intraocular pressure.
- *Diabetes:* The sugar level in the blood can lead to occlusion, and normally the blood sugar level is high in diabetic patients.
- *Smoking:* Heavy smoking habits may lead to occlusion.
- *Obesity:* People with heavy weight are always with the indication of more fat, and this content in the blood creates many health-related problems including high blood pressure and diabetes; this more probably creates occlusion.

In general pressure can optimize any body's capacity; similarly high pressure especially in diabetic people has impact on health consequences including occlusion for diabetic people, stroke, and some cardiac-related problems.

- *Stopping smoking:* The habit of smoking always reduces oxygen that reaches different parts of the body through the blood. Less oxygen flow harms the inner sides of blood vessels creating occlusion, so to avoid occlusion, people should stop their smoking habit.

- *Eating healthy food and controlled diet: Normally diseases can be controlled by eating habits and healthy diet.* To prevent from all the complications like occlusion and others, people should consume more fruits and vegetables; at the same time, they should not eat more fatty and oily food.
- *Avoiding or drinking a lesser amount of alcohol:* From the study of alcohol drinking people, it is found that alcohol and liquor progressively affect the neuron or veins related to organs in the body. But still there are many factors like age of person and lifestyle that have been considered while examining the effect.
- *Keeping active: Managing good health needs several practices including eating habits. A good and scheduled eating of food with good nutrients is required along with a good practice of sleep and hobbies. With all other good health practices, the person is required to walk early morning and make some physical exercises; these can keep the person active for a day.*
- *Managing good weight: Usually people having heavy weights used to have terrible eating habits and do not work more because of laziness. Since the food which is consumed by people remains with unused cholesterol and calories are not burn, this leads to accumulation of weight again.* People with increased eye pressure may likely to have a vessel occlusion. The optician can measure about high eye pressure and suggest the person to an ophthalmologist when required. *So to manage health retina and total health, it is important to have a balanced weight.*

There is a condition in the eye called glaucoma where eye weight causes problem to the optic nerve which is situated at the back of the eye. When the person is found with glaucoma situation, then it is very much essential to periodically monitor eye pressure and maintain it in control. Figure 8.4a–d shows different fundus cameras used for capturing retina image.

8.3.4   Blood Vessels

Blood vessels have a very crucial role in the functioning of the retina. All cells in the retina receive their food and nutrients from the blood which is supplied by the blood vessels. The main functions of the retina are to detect light incident and convert the light signals into electrical signals propagating them to the cerebrum part of the mind along the optic nerve. Once the cerebrum receives the electrical signals, then it converts these signals into an image that is being viewed. The healthy condition of blood veins or vessels is one of the crucial aspects, and normally the functioning of blood vessels is disturbed by diabetes disease. The pressure inside the blood vessels is one of the parameters that is to be controlled; usually people who have high sugar or diabetes disease tend to have normally more pressure inside the blood capillaries which makes the blood vessels weak because of pressure and sticky materials deposited in the inner side of blood veins which reduce size of veins by in turn creating again more pressure inside the blood capillaries.

The blood contains all nutrients which are required for any organs in the body to properly work. The blood flows normally through the veins, but there are enormous numbers of blood veins which carry blood to different parts of body. Similarly the nutrients to retina cells are also supplied by blood veins; even though their size is smaller, they do similar functions as other veins are doing.

Fig. 8.4 (**a**) Fixed fundus camera. (**b**) Portable fundus camera. (**c**) Handheld fundus camera. (**d**) Eye hospital ophthalmology fundus camera

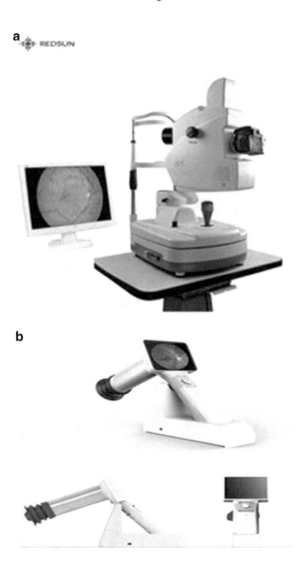

8.3.5 Exudates

There are some bright lesions inside the retina; among these exudates are main components. The exudates appearance is usually white patches, and these patches indicate severity of diabetes in the person. If the white patches are very less and situated in the first quadrant portion of the retina, then the disease is at the beginning stage. The white patches area occupied over retina surface is also one of the significant factors in making decision of diabetes presence. If the white patches are found at the first and second quadrants of retina surface area, then the diabetes has already made significant progress and requires immediate treatment. If exudates are detected in all four quadrants of retinal surface area, then it is the case of severity where immediate treatment is needed; otherwise there is threat of blindness. These exudates may be created at place, and they can have any shape. Sometimes large exudates may look like optic disc.

8.3.6 Microaneurysms

These are normally appearing much away from optic nerve where blood vessels originate and may be situated at place over the entire retina. Microaneurysms are primarily identifiable characteristics which appear as small red dots which contribute to local blood vessel damage.

Fig. 8.4 (continued)

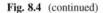

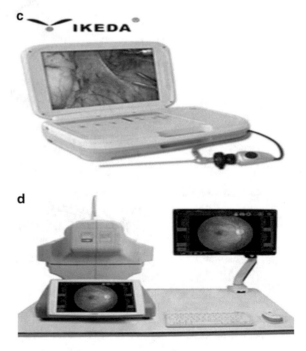

8.3.7 *Retinal Hemorrhage*

Hemorrhage is a disorder in which bleeding happens in the blood vessels through which blood is flowing to all the tissues of the retina, situated on the backside mass structure of the eye. There are photoreceptor cells in the retina called bars and cones; they can transfer light vitality into nerve flags that can be prepared by the mind to shape visual pictures. Retinal drain not exclusively can influence grown-ups, yet infants and babies may likewise suffer from this disorder. A retinal discharge can be brought about by a few ailments, for example, high pressure inside retinal blood vein impediment (there is a chance of blocking the blood vessels), sickliness, leukemia, or diabetes mellitus (which makes little delicate vein structures, which are effectively harmed).

Retinal hemorrhages that occur outside the macula can go undetected for a long time and may once in a while possibly be picked up when the eye is analyzed in detail by ophthalmoscopy, fundus photography, or an enlarged fundus test. However, some retinal hemorrhages can cause serious impairment of vision. They may happen regarding back vitreous detachment or retinal detachment. Figures 8.5, 8.6, and 8.7 show exudates details of different cases.

None: Exudates are not found.

Mild: Exudates area is less than 25% of total area of disc.

Moderate: Exudates area is than 25% of disc area and fewer than one disc area.

Severe: Exudates area is more than one disc area.

The following section describes the review in this area.

Fig. 8.5 Absence of exudates

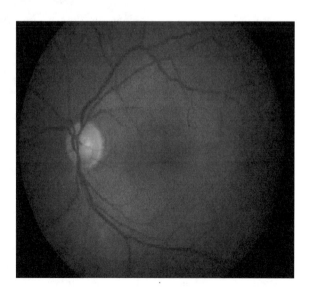

Fig. 8.6 Exudates presence

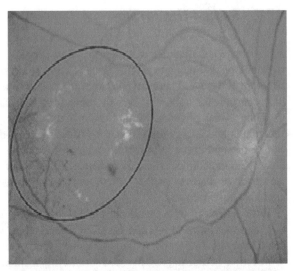

Fig. 8.7 High density of
exudates

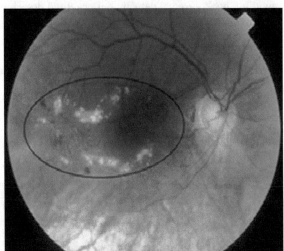

8.4 Literature Review

Retina images acquired by fundus camera are either uneven distribution of light or
small contrast posing tough challenges in the process of blood vessel localization
and segmentation. The following are some of the approaches made by different
authors.

Kuri [1] "presented a method to detect the blood vein network of the retina which
are the basic components of retinal structure. In this method they separated green
channel details from RGB image and primarily enhanced using "adaptive histogram
equalization technique," and further it is processed using Gabor filter along with

neighboring region entropy thresholding for mining blood vein network from retina image. The Gabor filter used in these sinusoids has modulated Gabor filter kernels. Local entropy threshold is done from the gray-level co-occurrence matrix.

Morales et al. [2] presented a scheme for differentiating diseased retina and healthy retina image. They extracted crucial features called local distribution of image binary pattern which serves as a strong gray-level surface feature. In this they generate a tag for every small detail in total image region. Label or tag generated will be used depending on image neighborhoods of the image pixels that are defined by means of radius and total number of points. In this the value of labels depends on the size of the neighborhood. Diverse binary patterns can be produced in each neighborhood.

Ali et al. [3] projected a method for analysis of diabetic retinopathy. They used a Gaussian low-pass filter to remove distortions and noise in the image. For segmentation purpose the histogram-based thresholding is used for eliminating the background of the retina image. The wavelet-based transform that uses Haar method is used in addition to Canny's periphery recognition and is performed to extract blood vessels.

Mansour [4] proposed a new process known as genetic algorithm along with vertex-based series cryptogram. A fractious position number and technique are applied on behalf of separation of the blood vein and intersection detection. This algorithm employs geometrical particulars of the retina blood vessels to diagnose hypertension and recognized retinal white spots known as exudates automatically from color retina images. The variation color intensity will make it difficult for detection of blood vessels.

Ramlogan et al. [5] employed the content of green channel of RGB image for effective recognition of retinal blood vessels. To enhance the retinal blood vessels adaptively, they used an enhancement technique known as "contrast limited adaptive histogram equalization" process over local regions. These large vessels are easily differentiated with respect to background since there exists a high intra-contrast. They divided the images into a block of size 8x8 partitions; furthermore, histograms of different detail equalization processes are operated on nonoverlapping relative regions. In this histogram equalization process, a small cutoff boundary is employed for small and thin veins. At the other end, a large cutoff boundary is considered in favor of small contrast difference along with respect to surroundings.

Youssef and Solouma [6] proposed a new technique to detect blood vessels of retina images; the method employed begins with edge recognition process. In this they used the latest feature vector-dependent approach for recognizing blood veins further more precisely. The method works with taking features of the blood vessels like intensity, width, length, and orientation for interested regions.

The segmented blood vessels containing optic disc and nerve are eliminated to obtain primary exudates. In the missed bright lesion, exudates are identified by applying morphological restoration process. This process has a difficulty in differentiating hard exudates and optic disc.

Cheng and Jhan [7] proposed the cascade-adaboost classifier. They applied support vector machine (SVM) classifier after that adaboost classifier to the same image regions and the image regions are known as "pedestrian candidate regions" with special camera was acquired for fixed-size images. The features were extracted from the above-said regions. They developed complete pedestrian classifier by joining SVM and adaboost classifiers in cascade. This reduced the rate of error of the system.

Dupas et al. [8] developed the algorithm to detect the microaneurysms and exudates from retinal images. They developed a system for determining computer-aided diabetes retinopathy and the classification of types of diabetes images. The accuracy of identification grading and of diabetes retinopathy (DR) and macular edema developed is further processed for improving performance of technique by considering big dataset containing all cases of disease and normal retina images.

Lam et al. [9] developed the local region's normalized contrast evaluation according to changes of special intensity in the image. For efficient blood vessel extraction process, perception of Weber's intensity law is applied for correlating the applied images to the system.

Joes et al. [10] proposed a technique to automatically separate the blood vessels from the image of a 2D color retina image obtained by fundus retina. This system is used for computerized screening of diabetic retinopathy (DR). The system focused around extraction of image edges that match very close to vessel centerlines. The edges provide preliminary information of line components, and images were approximated to patches by connecting every pixel in the image to its nearby line component. Every line component is used as a unique vector, and they were arranged using KNN classifier.

Yashawardhan et al. [11] presented another new method for blood vein local-ization in the fundus retinal images. They described blood vessel retinal image segmentation along with extraction of blood vessel method known as multilevel resolution of single channel and linear tracking. The processed images provide a remarkable view of the eye blood vessels that indicate the inspection of the blood vessels.

Jasprect et al. [12] presented computerized methodology for the purpose of detection and extraction of features from usual and abnormal abnormalities created in retinal images through employment of a filter-based method with a pool of Gabor filters for fragmenting the blood vessels. Further repetition and beginning of each Gabor channel have been adjusted to counterpart of the blood vessels that fall in that region of green channel content of input image since it contains dominant details of RGB retina image.

Safia et al. [13] established a comparative study involving two diverse techniques for diabetic retinopathy detection. The first method considered makes use of Gaus-sian shifting for preprocessing and log straightening out for division interpretation. The second method considered is employing of low concealing method in support of preprocessing the image; they applied "Gabor wavelet to upgrading in addition to global thresholding to segmentation."

Chaudhari et al. [14] presented a method for blood vessel segmentation. The method uses a 2D kernel having Gaussian, since a vessel can occur in any directions, the kernel is rotated in 12 different directions, and highest response in each pixel is retained designed for extraction of blood veins.

Lassada et al. [15] presented a technique to compute relationship of the edge localization and determining those edges which serve as blood vessels in the retina image. They proposed edge position policies for the purpose of automatic classification of retinal blood vessels in small child images. The method is tested and analyzed for accuracy of each technique with similarity of the results obtained by their method and manually graded ground-truth result.

Enrico et al. [16] have presented a new method for evaluation of segmented blood vessels from fundus images. The method is implemented such that the blood vessels are divided into small segments of constant curvature and later they have been combined to form complete vessels to find out true vessels.

Oliver et al. [17] reviewed detection of diabetes by processing retina images. The review contains many algorithms and techniques which are based on bright and red lesions. Mainly bright lesions are exudates which appear like white dots and optic disc which is circular in shape, whereas red lesions are microaneurysms and hemorrhages. At the end they discussed about classification based on different features extracted from the above-said lesions and entire retina images that can give characteristics of the retina.

Daniel et al. [18] developed retina video making where the process mainly concentrates on acquiring retina images from fundus images and later classifying them into diabetes and nondiabetes image cases from manual by two senior medical experts and computerized methods.

Vijayamadheswaran et al. [19] presented detection of diabetic retinopathy using radial basis function. The algorithm uses features obtained from the fundus images using contextual clustering (CC) segmentation method. The number of features obtained is two, and radial basis function (RBF) network is trained by the features; finally, weights are obtained and subsequently used for testing.

Gonzalez et al. [20] developed a new technique which is based on graph theory for technical analysis for segmentation and interpretation of blood vessels arrangement for retina image that estimates the location of blood vessels based on graph theory with intensity as a major information source and determined the unusual distribution of small and branched blood veins.

Kokare and Manjaramkar [21] developed a method to seek for microaneurysms which are mainly red dots. The algorithm starts with applying preprocessing method, namely, median filtering, and image histogram equalization later on the threshold is considered for differentiating background and retina lesions including red and white lesions. The threshold value is selected dynamically, and further they applied region growing method for segmentation and regression for fine classification.

8.4.1 Our Contribution

In this paper we proposed a new method for segmentation and classification of retinal images obtained with fundus camera. The proposed method consists of the following steps:

1. Preprocessing of input image that helps in segmentation process to distinguish different parameters of the retina.
2. Image segmentation is performed with the help of a two-level discrete wavelet transform along with mathematical morphology process to extract bright objects which are required in characterizing the images.
3. Two-level discrete signal-based wavelet transform has been employed to database image for obtaining wavelet characteristics also called features that serve as one feature vector.
4. Similar to step 3, wavelet features of the test are extracted that serve as a second set of feature vector.
5. Both the feature vectors computed in steps 3 and 4 are fed to three different distance measures considered, namely, city block, Minkowski, and Spearman.
6. The distance values generated by all three distance measures are tabulated.

Based on distance measure value, classification of images is performed. The method presented here provides effective accuracy of classification.

The part 1 of the paper provides introduction which is committed to describing about the diabetes disease types and spreading of disease in present days along with its consequences. This part also includes the characteristics along with literature which is committed to describing about the approach in which programmed processing techniques can be utilized for detection and scrutiny of diabetes.

The rest of the part of the chapter is organized as follows; later Sect. 2 provides a method developed for objective of segmentation and classification. In Sect. 3 results and discussion are provided which contain results obtained by proposed method that are tabulated and compared with already-existing methods. Section 4 gives conclusion drawn from the results obtained by method.

8.5 Proposed Methodology

In the classification of diabetes retina image machine learning classifiers are used, but many of the classifiers require training image set that should contain all possible types of diseased images, and it is very difficult to have all such cases; hence, we proposed a method that uses image similarity with respect to set of particular types such as normal, mild diabetic, moderate diabetic, and severe diabetic. Figure 8.8 indicates the building block diagram of the system developed.

Input image: The input image for the system is color fundus retina images for both database and test. The publically available retina image database diabet1 is used.

Fig. 8.8 Flowchart of proposed method

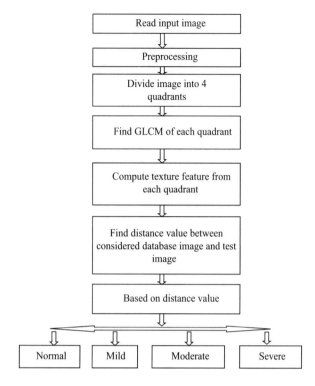

Preprocessing: In most of the cases, images contain noise and uneven illumination over its regions; therefore, preprocessing is needed. Adaptive method of histogram equalization is employed for removing noise.

The input image is a colored image; it converted gray-scale image that makes processing of image easy.

Segmentation: The preprocessed image is considered for segmentation. Mathematical morphology is applied to preprocessed image with a ball-shaped structuring element, and further discrete signal-based wavelet transform is performed to detect blood vessels. In this process discrete function f(n) is represented as a weighted summation of wavelets $\psi(n)$ and a common approximation $\varphi(n)$, and it is given by Eq. 8.1

$$f(n) = \frac{1}{\sqrt{M}} \sum_k W_\phi(j_0, E)\, \varphi_{j0,k} k(n) + \frac{1}{\sqrt{M}} \sum_{j=j_0}^{\infty} \sum_k W_\psi(j, k)\, \psi_{j,k}(n)$$

(8.1)

where the indices j_0 is an arbitrary starting value and $n = 0, 1, 2, 3, 4, 5 \ldots \ldots \ldots \ldots m$

Approximate coefficients are given by Eq. 8.2

$$w_\phi (j_0, k) = \frac{1}{\sqrt{M}} \sum_x f(x)\varphi_{j0,k}(x) \tag{8.2}$$

The detailed coefficients are given by Eq. 8.3

$$w_\phi (j_0, k) = \frac{1}{\sqrt{M}} \sum_x f(x)\varphi_{j0,k}(x) \tag{8.3}$$

The gray-level image is decomposed into four quadrants, and "gray-level co-occurrence matrix" (GLCM) is obtained for all quadrants. Texture features like contrast, homogeneity, energy, and correlations are calculated from every quadrant that yields a total of 16 features; along with these features, areas of blood veins of the retina are extracted, and they are used as another feature to form feature vector.

Contrast: The contrast can be a quantification of the neighboring variations in pixel values in an image and provides a good feature for further image representation, which is computed using Eq. 8.4.

$$C(k, n) = \sum_i \sum_j (i - j)^k \, P_d \, (i, j)^n \tag{8.4}$$

Homogeneity: The homogeneity of the image is co-occurrence of matrix of values with an amalgamation of high and low P [i, j] values in an image.

Entropy: Entropy is an evaluation of the information available in the image.

It evaluates the uncertainty of intensity distribution in the image, and it is represented by using Eq. 8.5.

$$C_e = \sum_i \sum_j P_d \, (i, j) \ln \, P_d \, (i, j) \tag{8.5}$$

Correlation: Eq. 8.6 is used to measure the correlation of image linearity.

$$C_e = \frac{\sum_i \sum_j [i \, j \, P_d \, (i, j)] - \mu_i \mu_j}{\mu_i \mu_j} \tag{8.6}$$

where $\sigma^2 = \sum i^2 P_d \, (i,j) - \mu_i^2$
If the correlation is higher, then the image consists of large amount of linear structure.

Energy: One of the ways for generating texture features is to utilize local kernels. After the multiplication and summing of pixels with the particular kernel at all image coordinates, the "texture energy measure" (TEM) is found by adding the absolute values in a local region. The energy is computed using Eq. 8.7.

$$L_e = \sum_{i=1}^{m} \sum_{j=1}^{n} |C(i, j)| \tag{8.7}$$

8.5.1 Disease Classification

Changes in the retina caused by diabetes can be classified as four different classes including healthy or normal class, mild class, moderate class, and severe class. The normal or nondiseased retina image does not contain any white patches known as exudates, red spots known as macula, or hemorrhages, and blood vessels are not dilated or thickened. The mild diabetes retina image contains blood vessels dilated and one or two exudates in the entire image area. The moderate diabetic retina image contains exudates in the first and second quadrants of the image, and the exudates area is significantly more along with exudates blood vessels which are also dilated. In severe diabetes retina image, exudates are distributed in all four quadrants of the image with presence of dilated blood vessels, red spots, macula, and hemorrhages. Figure 8.9a, b, c shows different stages of diabetes progress.

8.5.2 Datasets

The database used in the testing of this method is taken from an online publically available source. The total images available in this database are 89, and all are color fundus images; among them 84 images present at least small nonproliferative cipher of the diabetes and 5 images show normal characteristics where there are no or any symbols of diabetes, as per the experts involved in the evaluation. The images

Fig. 8.9 (**a**) Mild diseased retina. (**b**) Moderate diseased retina. (**c**) Severe diseased retina

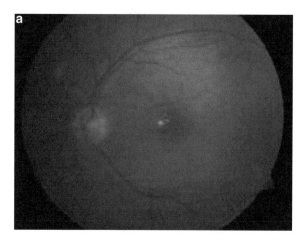

Fig. 8.9 (continued)

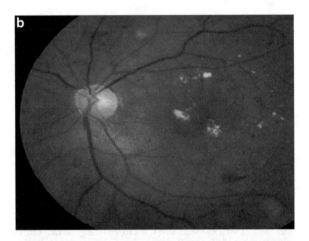

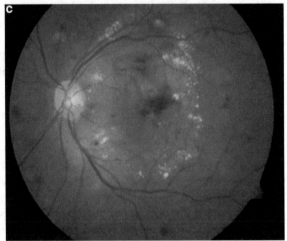

are collected from "Kuopio University Hospital." The images of database used are acquired with digital fundus camera having a 50-degree field view. Figure 8.9a–c shows possible diseased cases of retina images.

8.5.3 Distance Measures

The input image is M × N data matrix known as X matrix, which is considered as 1 × N row vectors, $x_1, x_2,..., \ldots\ldots\ldots x_m$.

At the same time, M × N data matrix is known as Y matrix, which is considered as 1 × N row vectors, $y_1, y_2, ..., y_m$.

Table 8.1 Comparisons of distance measure performance

Reference images	No. of database images	Type of image	No. of test images	Avg. distance (city block)	Avg. distance (Minkowski)	Avg. distance (Spearman)
Normal	10	Normal	10	0.150	0.618	0.0170
		Mild	10	0. 224	0.0637	0.0190
		Moderate	10	0.314	0.0851	0.0209
		Severe	10	0.554	0.1341	0.0364

Then city block distance, Minkowski distance, and Spearman distance are computed between the vectors x_s and y_t using Eqs. 8.8, 8.9, and 8.9 correspondingly.

$$D^2 = (x_s - y_t)(x_s - y_t)^1 \tag{8.8}$$

$$D^2 = \sum_{j=1}^{n} |x_{sj} - y_{tj}| \tag{8.9}$$

$$D = 1 - \frac{(r_s - r^-_s)(r_t - r^-_t)^1}{\sqrt{(r_s - r^-_s)(r_t - r^-_s)^1}\sqrt{(r_t - r^-_t)(r_t - r^-_t)^1}} \tag{8.10}$$

Table 8.1 shows numerical values that are obtained for finding out threshold distance value of three distance measures that can be used as a decision-making rule to classify all images of database.

8.6   Results and Discussion

The database and test images are selected from diabet1 database since it contains all grades of diabetic images such as mild, moderate, and severe along with some normal images.

The similarity measures like city block and Minkowski methods are used to find the similarity content of two images. A set of ten different normal images from database are considered and called reference images; again another set of ten normal, mild, moderate, and severe DR are compared with the help of distance measures; and similarity value is tabulated in Table 8.1, which helps in the classification of images. Here reference images can be normal, mild, moderate, or severe DR images. This method gives good result because all images of database are similar even though they belong to the same type. Similar types of retina image when compared by distance measures should minimum value and dissimilar images should give maximum value. Among these three different distance measures used, city block distance measure gives distinguishing values for all three combinations of images as used in Table 8.1, like normal against normal, normal against mild, normal against moderate, and normal against severe.

On the bases of distance value obtained, it is decided whether the test image belongs to normal, mild, moderate, or severe case. The performance measure of the proposed method is given in Table 8.2, which contains classification accuracy.

For computing performance of the method, the database diabet1 is considered, and each image from the database is compared with the dataset of ten normal images which are used as reference images. City block distance measure to find similarity. Figure 8.10 shows distance measure values of all three distance measures, and Fig. 8.11 shows the classification accuracy of the system.

Table 8.2 Classification accuracy

Methods	Classification accuracy
K. Narasimhan et al. [16]	70
Meindert Niemeijer et al. [17]	80
Proposed	91.10

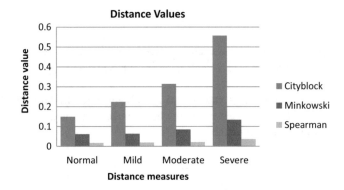

Fig. 8.10 Distance values versus distance measures

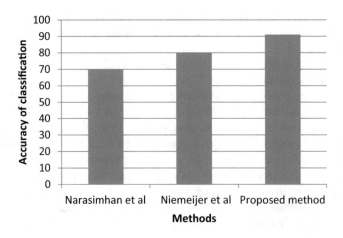

Fig. 8.11 Classification accuracy

8.7 Conclusion

Similarity measures are mainly used when two images are to be compared for their similarity content in them." In the proposed method, we used three similarity measures, city block, Minkowski, and Spearman, for finding the similarity between two images; all three methods clearly distinguish the different types of images like normal, mild, moderate, and severe from a group of images by providing different ranges of distance values which indicate the category of retina image." Most of the classifiers require large training dataset for their classification, and if the training set does not contain all categories of images, then the test of classification will be erroneous, since this method requires only one type of image as reference and test will be more robust over other disease classifiers.

References

1. S.K. Kuri, Automatic diabetic retinopathy detection using Gabor filter with local entropy thresholding, Recent Trends in Information Systems ReITS-2015, 2nd IEEE International Conference on, pp. 411–415, 2015
2. S. Morales, K. Engan, V. Naranjo, A. Colomer, Detection on diabetic retinopathy" and age related macular degeneration from fundus images through local binary pattern and random forests. ICIP -2015. IEEE International Conference, pp. 4838–4842, 2015
3. A. Shojaeipour, Md.J. Nordin, N. Hadavi, Using image processing methods for diagnosis of diabetic retinopathy, ISRMA-2014, IEEE International Conference on, pp. 154–159, 2014
4. R.F. Mansour, Using genetic algorithm for identification of diabetic retinal exudates in digital color fundus images. J. Intell. Learn. Syst. Appl. **04**, 188–198 (2012)
5. G.S. Ramlugun, V.K. Nagarajan, C. Chakrabarty, Small retinal vessels extraction towards proliferative diabetic retinopathy screening. Expert Syst. Appl. **39**, 1141–1146 (2012)
6. D. Youssef, N.H. Solouma, Accurate detection of blood vessels improves the detection of exudates in color fundus images. Comput. Methods Prog. Biomed. **108**, 1052–1061 (2012)
7. C. Cheng, D.M. Jhan, A cascade classifier using Ada-boost algorithm and support vector machine for pedestrian detection, IEEE International conference on Systems Man and Cybernetics(SMC)-2011, pp. 1430–1435, 2011
8. B. Dupas, T. Walter, A. Erginary, R. Ordonez, N.D. Joardar, P. Gain, Evaluation of automated fundus photograph analysis algorithm for detecting microaneurysms, haemorrhages and exudates, and of a computer assisted diagnostic system for grading diabetic retinopathy. Journal of diabetes and metabolism **36**, 213–220 (2010)
9. B.S. Lam, Y. Gao, A.C. Liew, General retina vessel segmentation using regularization-based multi concavity modeling. IEEE Trans. Med. Imaging **29**, 1369 (2010)
10. S. Joshi, P.T.K. Dr, Retinal blood vessel segmentation. Int. J. Eng. Innov. Technol. (IJEIT) **1**(3), 175–178
11. K. Yashowardhan, I.K. Mohammed, "A newly rule-based method for blood vessel segmentation" and "multi-scale single-channel linear tracking method". Int. J. Comput. Technol. Appl. **4**(2), 369–373
12. K. Jaspreet, H.P.S. Dr, Automated detection of retinal blood vessels in diabetic retinopathy using Gabor filter. Int. J. Comput. Sci. Netw. Secur. **12**(4), 109–116 (2012)
13. S. Safia, T. Anam, M.U. A, A comparison and evaluation of computerized methods for blood vessel enhancement and segmentation in retinal images. Int. J. Future Comput. Commun. **2**(6), 600–603 (2013)

14. S. Chaudhari, S. Chattergee, N. Katz, M. Nelson, M. Goldbaum, Detection of blood vessels in retinal images using 2D matched filters. IEEE Trans. Med. Imaging **8**(3), 263–269 (1989)
15. S. Lassada, U. Bunyarit, B. Sarah, Comparison of Edge Detection Techniques on Vessel Detection of Infant's Retinal Image, Proceedings of the International Conference on Computer and Industrial Management, ICIM, October 29–30, 2005, Bangkok, Thailand, pp 61–65, 2005
16. E. Grisam, M. Foracchia, A. Ruggeri, A novel method for automatic grading of retinal vessel tortuosity. IEEE Trans. Med. Imaging, 1–13 (2007)
17. O. Faust, U.R. Acharya, E.Y. Ng, K.-H. Ng, J.S. Suri, Algorithms for the automated detection of diabetic retinopathy using digital fundus images: A review, *Springer science and business media LLC*. J. Med. Syst. **36**(4), 145–157 (2010)
18. D.S.W. Ting, M.L. Tay-Kearney, I.C. Franzco, L.L. Franzco, D.B.P. Franzco, Y. Kanagasingam, Retinal Video Recording, in *American Academy of Ophthalmology*, (Elsevier Inc, 2011), pp. 1588–1593
19. M.R. Vijayamadheswaran, D.M. Arthanari, M.M. Sivakumar, Detection of diabetic retinopathy using radial basis function. Int. J. Innov. Technol. Creat. Eng. **1**(1), 40–47 (2011)
20. A. Salazar-Gonzalez, D. Kaba, Y. Li, X. Liu, Segmentation of the blood vessels and optic disk in retinal images. IEEE J. Biomed. Health Inform. **18**(6), 1874 (2014)
21. M. Kokare, A. Manjaramkar, Decision Trees for Microaneurysms Detection in Color Fundus Images, IEEE International Conference on Innovations in Green Energy and Healthcare Technologies (ICIGEHT'17), 2017

Chapter 9
Language-Based Classification of Document Images Using Hybrid Texture Features

Umesh D. Dixit and M. S. Shirdhonkar

9.1 Introduction

Rapid development of technology has given rise in number of document images in every language. A huge database of multilingual document images needs an automatic language-based classification system. The language-based classification of documents has following applications:

- Document categorization based on the domain
- Retrieving of documents
- OCR implementation
- Digital libraries
- Text to speech conversion

Figure 9.1 shows taxonomy of language-based/script-based classification of document images. Both global and local analysis can be applied to identify the scripts. Global analysis includes feature extraction at paragraph or block level. Local analysis of document shall be implemented at two levels: word level and line level. In the word level, initially the document image is segmented into words using connected component analysis. Normally the structural features, texture features, or hybrid features are employed. Structural features consist of information about character strokes, orientation, and their sizes. Texture features are the presentation of visual appearance of components and their frequency. Hybrid features include the usage of both structural and texture features. In the word-level implementation,

U. D. Dixit (✉)
Department of Electronics & Communication Engineering, B.L.D.E.A's, V.P. Dr. P.G. Halakatti College of Engineering & Technology, Vijayapur, Karnataka, India

M. S. Shirdhonkar
Department of Computer Science & Engineering, B.L.D.E.A's, V.P. Dr. P.G. Halakatti College of Engineering & Technology, Vijayapur, Karnataka, India

© Springer Nature Switzerland AG 2019
G.R. Sinha (ed.), *Advances in Biometrics*,
https://doi.org/10.1007/978-3-030-30436-2_9

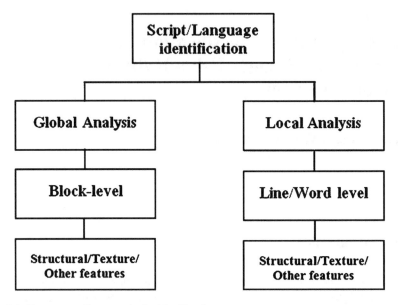

Fig. 9.1 Taxonomy of language/script identification

the extracted features of the connected components are employed for classification of the document images. The line-level implementation includes segmentation of lines using horizontal and vertical profiles of the text, feature extraction, and then classification of the documents based on the language. However in paragraph or block-level implementation entire document image is treated as a block. The features from this block are extracted to train the system in order to carry out language-based document image classification.

As the global analysis of document image is segmentation-free, it has an advantage of faster classification. The local analysis requires an additional preprocessing time for segmentation of the document, but has an advantage of accurate classification. To improve both speed and accuracy of classification, there is a need for development of new feature extraction schemes.

The objective of the work presented in this chapter is to develop a system that classifies printed document images based on the language used. This chapter proposes a segmentation-free technique for language-based classification. The proposed method is evaluated for Kannada, Telugu, Marathi, Hindi, and English documents. Kannada, Telugu, and Marathi are the official languages of Karnataka, Telangana, and Maharashtra states of India. Hindi and English are the national and global languages, which are officially accepted across India. Figure 9.2 shows sample document images of Kannada, Telugu, Marathi, Hindi, and English.

Kannada and Telugu scripts are derived from Brahmi alphabet of ancient India. During the twelfth and fifteenth century, these two scripts are split into separate alphabets. The Kannada language has 16 vowels and 34 consonants with 250 basic, compound, and modified shapes. The Telugu language includes 16 vowels, 3 vowel modifiers, and 41 consonants. Thus it has a total of 60 symbols. Hindi and Marathi

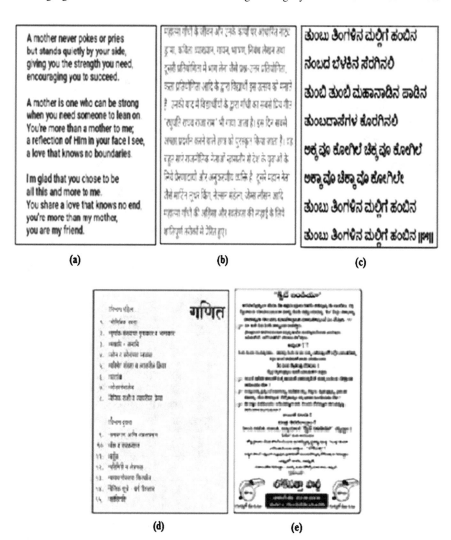

Fig. 9.2 Sample documents of English, Hindi, Kannada, Marathi, and Telugu

are derived from Devanagari script. Both Hindi and Marathi scripts have a horizontal line at the top and connects all the letters. The Hindi script has 12 vowels and 34 consonants, whereas Marathi includes 16 vowels and 36 consonants. English uses Latin-based alphabets with 26 letters. It has 5 vowels and 21 consonants. The important feature of English alphabets is that most of them have vertical and slant strokes. Figure 9.3 shows the vowels and consonants of Kannada, Telugu, Marathi, Hindi, and English languages.

The challenging task in the proposed document classification is due to similarity between Hindi and Marathi scripts as well as Kannada and Telugu scripts. But the texture features formed by the words used in these languages will be distinct. This

Fig. 9.3 Vowels and consonants of Kannada, Telugu, Marathi, and Hindi languages. (**a**) Kannada vowels and consonants. (**b**) Telugu vowels and consonants. (**c**) Marathi vowels and consonants. (**d**) Hindi vowels and consonants. (**e**) English vowels and consonants

(a)

(b)

Initial Vowels

a	ā	i	ī	u	ū	ṛ	e	ai	o	au
[ə]	[a]	[i]	[i:]	[u]	[u:]	[ri]	[e]	[ai]	[o]	[ɔu]

Velar and Palatal

ka	kha	ga	gha	ṅa	ca	cha	ja	jha	na
[kə]	[kʰə]	[gə]	[gʰə]	[nə]	[tsə/tʃə]	[tsə/tʃʰə]	[dzə/dʒə]	[dzʰə/dʒʰə]	[nə]

Retroflex and Dental

ṭa	ṭha	ḍa	ḍha	ṇa	ta	tha	da	dha	na
[ʈə]	[ʈʰə]	[ɖə]	[ɖʰə]	[ɳə]	[tə]	[tʰə]	[də]	[dʰə]	[nə]

Labial and Semivowel

pa	pha	ba	bha	ma	ya	ra	la	va
[pə]	[pʰə]	[bə]	[bʰə]	[mə]	[jə]	[rə]	[lə]	[va/wə]

Fricative, Retroflex Liquid and biconsonantal groups

śa	ṣa	sa	ha	ḷa	kṣa	jña	śra
[ʃə]	[ʃə]	[sə]	[ɦə]	[lə]	[kʃə]	[jnə]	[ʃrə]

(c)

Fig. 9.3 (continued)

(d)

Vowels:

A E I O U

Consonants:

B C D F G H J K L M N P Q
R S T U V W X Y Z

Vowels:

a e i o u

Consonants:

b c d e f g h i j k l m n o p q
r s t u v w x y z

(e)

motivated us to propose a suitable texture features for language-based classification of document images. The important contribution of this chapter is proposing the usage of hybrid features employing SWT and HOG to improve classification accuracy. The chapter also presents comparative analysis of proposed hybrid features with (1) rotation invariant local binary pattern, (2) histogram of oriented gradients, and (3) multi-resolution HOG feature. Evaluation of the method is carried out on a database of 1006 document images. The proposed feature extraction scheme with SVM classifier provided better classification accuracy compared with current state-of-art techniques.

The rest of the chapter is organized as follows:

- Section 9.2 briefs the literature review of language-based identification of document images.
- Section 9.3 details about the methodology.
- Section 9.4 presents experimental results and Sect. 9.5 concludes about the work.

9.2 Literature Review

Script/language identification is a subfield of document image analysis. Lot of work has been carried out in document image analysis. A detailed survey on document image analysis is provided in [1]. The following section deals with the work related to script/language identification.

Chaudhury et al. [2] presented script identification system for Indian languages. They used Gabor filter-based features extracted from connected components, with combined classifiers to improve the performance. Kulkarni et al. [3] proposed script identification from multilingual documents using visual clue-based features. Eight different visible features are employed with probabilistic neural network (PNN). Padma and Vijaya [4] employed profile-based features with k-nearest neighbor classifier for script identification from trilingual documents.

Pal and Chaudhuri [5] developed a system for identification of English, Bangla, Arabic, Chinese, and Devanagari script lines from a document. They combined shape-based, statistical-based, and some of the water reservoir-based features in their work. Rajput et al. [6] presented a system for handwritten text identification using DCT and the wavelet features. They processed document at block level and used K-nearest neighbor approach for classification. Mathematical and structure-based features with a series of classifiers have been applied to improve the performance of script identification for Indian document images in [7]. Shirdhonkar and Kokare [8] presented a technique to discriminate printed text and handwritten text using neural network model and SVM.

Pardeshi et al. [9] used multi-resolution spatial features for Indian script identification. They extracted features by applying radon transform, DWT, and DCT on the segmented words of the document images. Tan et al. [10] used word shape analysis to retrieve text from the document images. Wanchoo et al. [11] provided a survey of Devanagari script recognition for Indian postal system. Sahare and Dhok [12] detailed about the algorithms used for recognition of text in their work. They also included comparison of the different schemes used for text recognition. A detailed survey on document image analysis with its applications, challenges, and current state of art is presented by Dixit and Shirdhonkar [13].

Arani et al. [14] used hidden Markov model (HMM) for recognition of handwritten Farsi words. They used multilayer perceptron (MLP) with an input of features obtained from image gradient, contour chain code, and black-white transitions. Bi et al. [15] presented their final version of Chinese handwritten character recognition system using convolutional neural network (CNN) model with GoogLeNet. Djeddi et al. [16] presented a system for writer recognition using multi-script handwritten text, comprising of Greek and English languages. They used run length features with k-NN and SVM classifiers. Dixit and Shirdhonkar [17] proposed the multi-resolution LBP features in their fingerprint-based document image retrieval work. They compared the results of multi-resolution LBP features obtained using DWT and SWT. Roy et al. [18] proposed HMM-based Indic handwritten word recognition system. They used features obtained from zone-wise segmentation of words in their work.

In the literature, we found that most of the text recognition schemes are based on the line level or word level. Only a handful of works were reported text recognition at block level or document level. Segmentation-free script identification at block level improves the classification speed and is helpful to retrieve documents based on the script. The scripts used to form words and sentences, in different languages, will have visually distinct features. We exploited these features for language-based classification. A hybrid feature extraction scheme that combines SWT and HOG is proposed for improved performance. We used SWT to decompose the document image into horizontal, vertical, and diagonal details of the scripts. The decomposition helped in obtaining more precise orientation of gradients to construct the feature vector.

9.3 Proposed Methodology

Figure 9.4 provides architecture of the proposed work. It includes training phase and testing phase. Preprocessing, feature extraction, and classification are the building blocks of the proposed architecture. These blocks are discussed in the subsequent sections.

9.3.1 Preprocessing

In this step the document image is prepared for feature extraction. Initially the document image is converted into grayscale using the equation (9.1):

$$I = 0.2989 \times R + 0.5870 \times G + 0.114 \times B \tag{9.1}$$

Fig. 9.4 Proposed
language-based classification

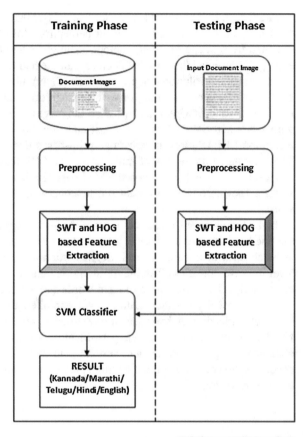

Fig. 9.5 Unsharp filter mask

1	-1	-1
-1	8	-1
-1	-1	-1

"R," "G," and "B" in the equation (9.1) are the red, green, and blue components of the image. The converted grayscale image is denoted as "I." Low-contrast document images provide inaccurate texture features and lead to poor classification. Hence to increase quality of the input image for improved classification, we convolved the image "I" with a 3 × 3 filter shown in Fig. 9.5 to perform unsharp masking [19] and then applied a low-pass filter.

Figure 9.6 shows the results of preprocessing steps for a sample document image. It includes input color image, its grayscale version, output of unsharp masking, and the low-pass filtered image. Thus in the preprocessing step, we improve quality of the document image in terms of contrast to obtain more accurate texture features in the next step. The steps used in preprocessing of the document image are listed in Algorithm 9.1.

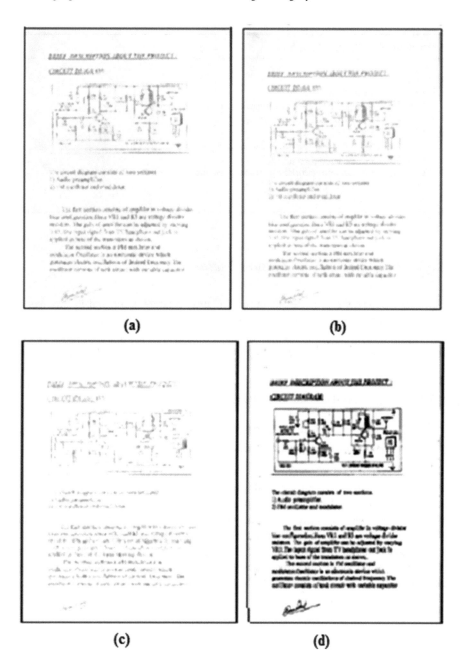

Fig. 9.6 Results of preprocessing steps

Algorithm 9.1 Preprocessing

```
1. Begin
   Input: Document image
   Output: Preprocessed document image D(x,y).
2. Read the input document image
3. if input is color image
       Convert to gray-scale using equation (9.1)
   end if
4. Perform un-sharp masking using mask shown in Fig. 9.5
5. Apply low-pass filter
6. End
```

9.3.2 Proposed Hybrid Texture Features

Figure 9.7 shows the proposed feature extraction scheme employed in this work. We used SWT- and HOG-based hybrid features. The document image is initially decomposed by applying DWT and then HOG features are obtained from each of these decomposed components. This process is explained in the following sections.

9.3.2.1 Stationary Wavelet Transform (SWT)

This section provides the details of DWT, SWT, and the application of SWT in the proposed method of language-based document image classification.

DWT The word wavelet was first introduced by Morlet and Grossman in the design of Morlet wavelet. In 1984, wavelet with new property called orthogonality was proposed by Meyer. The orthogonal property states that the information obtained by one wavelet will be entirely independent of the information captured by another wavelet. An idea of multi-resolution that is a pyramidal algorithm was developed by Stephane Mallat in 1986. The kernel functions used in wavelet transform are obtained by a prototype function referred to as mother wavelet, which is given by equation (9.2):

$$\psi_{a,b}(t) = \frac{1}{\sqrt{a}} \psi \frac{t-b}{a} \tag{9.2}$$

where "a" is the scaling factor and "$t - b$" is the translation parameter. The term $1/\sqrt{a}$ is used as normalization factor to ensure that all the wavelets carry same energy. Thus the wavelet of a signal is a mother wavelet at scale "a," lagged by b. Figure 9.8 shows plot of a mother wavelet.

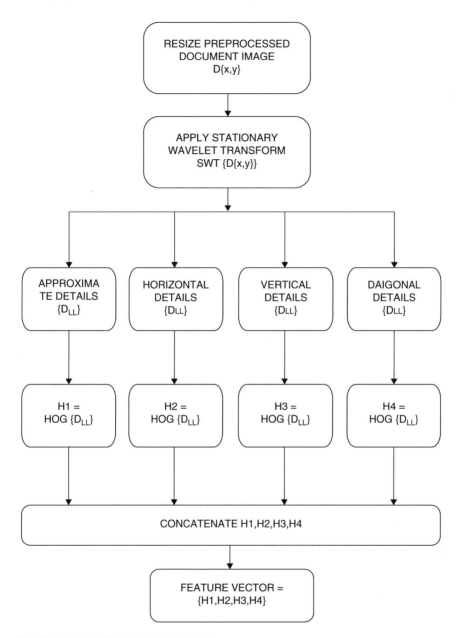

Fig. 9.7 Proposed feature extraction scheme

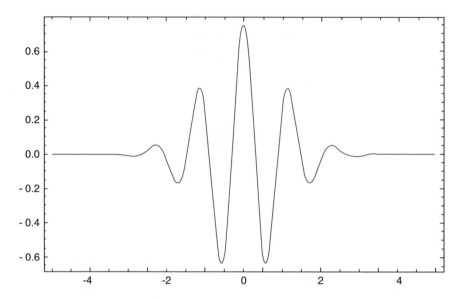

Fig. 9.8 Plot of a mother wavelet

The discrete wavelet transform provides discrete samples of a wavelet transform. DWT of a 2D signal $f(x,y)$ is given by equations (9.3) and (9.4):

$$W_\varnothing (j_O, m, n) = \frac{1}{\sqrt{M \times N}} \sum_{x=0}^{M-1} \sum_{y=0}^{N-1} f (x, y) \, \varnothing_{j_O, m, n} (x, y) \qquad (9.3)$$

$$W_{\psi}^i (j_O, m, n) = \frac{1}{\sqrt{M \times N}} \sum_{x=0}^{M-1} \sum_{y=0}^{N-1} f(x, y) \psi_{j_O, m, n}^i (x, y) \qquad (9.4)$$

where:

- j_0 is arbitrary starting scale.
- $W_\varnothing(j_0, m, n)$ is an approximation of $f(x,y)$ with scale j_0.
- $W_{\psi}^i(j, m, n)$ represents horizontal, vertical, as well as diagonal details with scale $j \geq j_0$.
- M and N are row and column dimensions of the input image.

Figure 9.9 shows the conceptual approach of obtaining two-dimensional DWT, which is obtained using the series of low-pass and high-pass filters [20, 21]. The notations "H" and "G" correspond to low-pass and high-pass filters, obtained by convolving the image with moving average and moving difference masks. The ②↓ and ①↓ indicate down-sampling of columns and rows, respectively. The result of this operation leads to four sub-bands:

Fig. 9.9 DWT
decomposition of an image

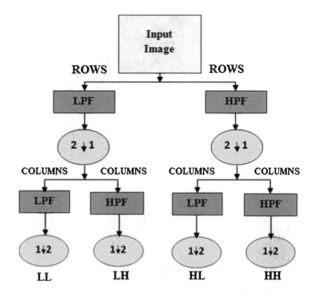

- LL – Approximate sub-band that contains down-sampled original image
- LH – Horizontal details of an input image
- HL – Vertical details of an input image
- HH – Diagonal details of an input image

The DWT can also be applied at multilevels to obtain more accurate features. Figure 9.10 shows the result of applying 2D DWT on an image with approximate, horizontal, vertical, and diagonal details.

The two-dimensional DWT is found to be useful in many image processing applications, which include:

- Image compression
- Image denoising
- Steganography
- Feature extraction
- Turbulence analysis
- Topographic data analysis
- Financial analysis and many more

Stationary Wavelet Transform (SWT) The discrete wavelet transform lacks with translation invariance property. The SWT is designed to provide translation invariance features and is an improved version of DWT. Translation invariance is achieved by eliminating up-sampling and down-sampling process, followed by up-sampling the filter coefficients with 2^{J-1} factor in J^{th} level of discrete wavelet transform (DWT) algorithm [22–27]. Figure 9.11 shows decomposition of the image using SWT. It can be observed that the up-sampling and down-sampling are removed from the process. The LL, LH, HL, and HH are decomposed components of the input image. In SWT, the number of output samples in every level of decomposition is same as that of input samples.

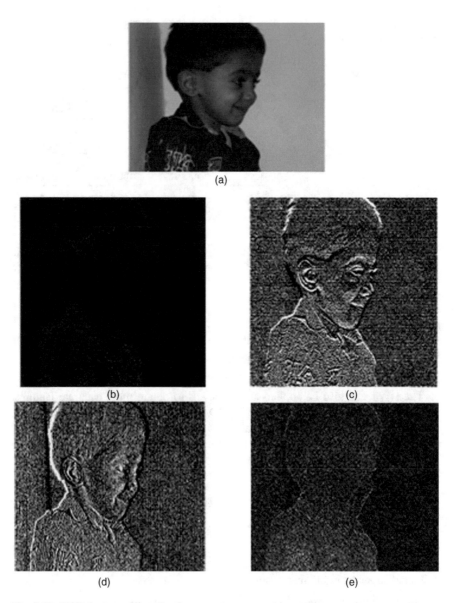

Fig. 9.10 DWT decomposition of an image

In the proposed method, the preprocessed document image is resized to 256×256 pixels. Let $D(x,y)$ be a resized document image. Applying SWT on $D(x,y)$ produces four sub-bands of image, namely, D_{LL}, D_{LH}, D_{HL}, and D_{HH}, as given by equation (9.5):

$$\text{SWT}\{D(x, y)\} = \{D_{LL}, D_{LH}, D_{HL}, D_{HH}\} \qquad (9.5)$$

Fig. 9.11 SWT decomposition of the image

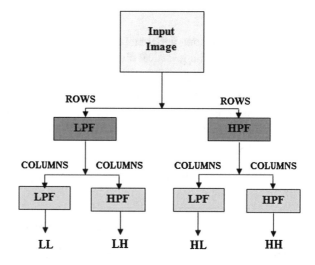

where D_{LL} contains approximation coefficients, D_{LH} includes horizontal coefficients, D_{HL} consists of vertical coefficients, and D_{HH} contains diagonal coefficients. Thus the four sub-bands provide multi-resolution version of the input document image with translation invariance features. This decomposition helps in acquiring more precise features in the next step.

9.3.2.2 Histogram of Oriented Gradients (HOG)

HOG was initially proposed by Dalal and Triggs [28] for human face detection. Later HOG and its variants are used for hand detection [29], pedestrian detection [30], fast face recognition, [31] and in many more image recognition applications.

In the proposed work, the four decomposed versions of the image obtained using SWT are divided into blocks of size 2×2 cells. This gives a total of 16 cells per image. Deciding size of the cell is an important step during extraction of HOG features. We tested the system with cells of size 128×128 and also 64×64 pixels. Cells of size 128×128 yielded a feature vector of dimension 144 and cells of size 64×64 yielded 1296 features. In our experiments we found that 128×128 cells with 144 features provided better results in comparison with 64×64 cell size.

After dividing the blocks into cells, the gradient and orientation of the pixels in each cell is computed. The gradient of the pixels is first order derivative and it gives finer details of the image. The gradient of a 2D function $f(x,y)$ is a column vector represented using equation (9.6):

$$\begin{bmatrix} \nabla x \\ \nabla y \end{bmatrix} = \begin{bmatrix} \frac{\partial f}{\partial x} \\ \frac{\partial f}{\partial y} \end{bmatrix} = \begin{bmatrix} H_G(x, y) \\ V_G(x, y) \end{bmatrix} \tag{9.6}$$

where:

- ∇x is gradient along horizontal direction and is represented as $H_G(x,y)$.
- ∇y is gradient along vertical direction and is represented as $V_G(x,y)$.

However, in image processing the equations (9.7) and (9.8) are used to compute the gradient along horizontal and vertical directions of pixels in each cell. The gradient in horizontal direction $H_G(x,y)$ is difference of two successive pixels of a row and the gradient in vertical direction $V_G(x,y)$ is difference of two successive pixels of a column:

$$H_G(x, y) = D_i(x + 1, y) - D_i(x - 1, y) \tag{9.7}$$

$$V_G(x, y) = D_i(x, y + 1) - D_i(x, y - 1) \tag{9.8}$$

The magnitude Mag(x,y) and the direction of gradients $\Theta(x,y)$ are obtained using equations (9.9) and (9.10). The magnitude represents strength of the edge point and the direction gives orientation of the pixel at location (x,y):

$$\text{Mag}(x, y) = \sqrt{H_G(x, y)^2 + V_G(x, y)^2} \tag{9.9}$$

$$\Theta(x, y) = \tan^{-1} \frac{G_H(x, y)}{G_V(x, y)} \tag{9.10}$$

In the next step, the orientation of gradient value of the pixels from each cell is represented as a histogram. Histogram is a discrete function that provides information about number of occurrences of a specific data. The histograms obtained for each sub-band of the image are concatenated to form final set of features. Let H_1, H_2, H_3, and H_4 be the histograms of sub-bands D_{LL}, D_{LH}, D_{HL}, and D_{HH}, respectively. These four histograms are concatenated to form a final set of features for language-based classification. Let FV be a final feature vector obtained using equation (9.11):

$$\text{FV} = \{H(D_{LL}) \ U \ H(D_{LH}) \ U \ H(D_{HL}) \ U \ H(D_{HH})\} \tag{9.11}$$

The range of values used in a histogram is called bins. We used 9 bins to store frequency of gradient values for each cell. As each sub-band comprises of four cells, we get $9 \times 4 = 36$ features per sub-band image. Thus a total of $36 \times 4 = 144$ features per document image are obtained for classification.

Figure 9.12 shows SWT decomposition of the sample document image with horizontal, vertical, and diagonal coefficients. Figure 9.13 depicts the plot of proposed feature values in graphical form. Algorithm 9.2 enlists the steps adopted in proposed feature extraction scheme.

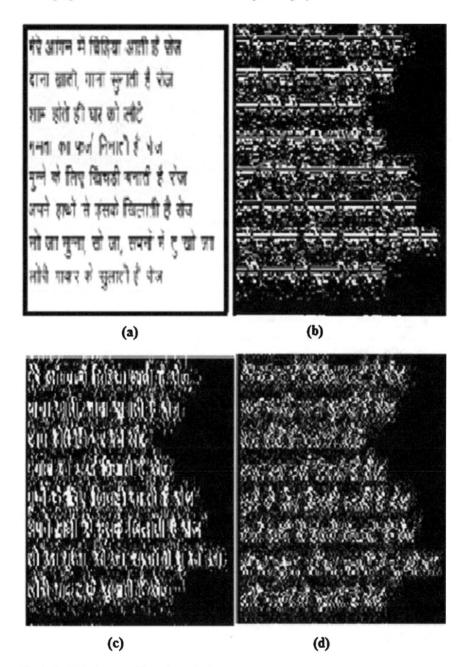

Fig. 9.12 SWT decomposition of sample document

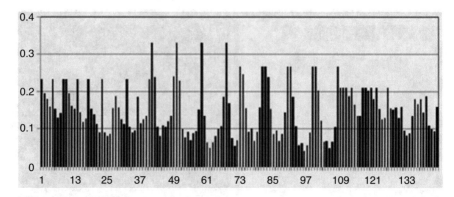

Fig. 9.13 Plot of feature values for sample document image

Algorithm 9.2

Proposed feature extraction scheme

```
1. Begin
   Input: Pre-processed document image D(x,y)
   Output: Feature vector (FV)
2. Resize the document image D(x,y) to 256×256 pixels.
3. Apply stationary wavelet transform on the image D(x,y).
   [D_LL, D_LH, D_HL, D_HH] = SWT {D(x,y)}
4. Extract HOG features from D_LL, D_LH, D_HL and D_HH.
   (a) H_1 = HOG (D_LL)
   (b) H_2 = HOG (D_HL)
   (c) H_3 = HOG (D_LH)
   (d) H_4 = HOG (D_HH)
5. Concatenate H1, H2, H3 and H4 to construct feature vector.
6. FV = { H_1 U H_2 U H_3 U H_4 }
7. End
```

9.3.3 SVM Classifier

SVM belongs to the supervised machine learning technique and is widely found in the application of image classification [32, 33]. The advantages of SVM are:

- It is effective in high-dimensional space.
- It provides better results with less number of test samples.
- It is more versatile, due to a large number of kernel functions.

As SVM is a supervised learning, it requires training with some known data. Classification of the test data in SVM is obtained using an optimal hyperplane. This line separates the data into two classes. An example of separating circles and squares with a hyperplane is shown in Fig. 9.14.

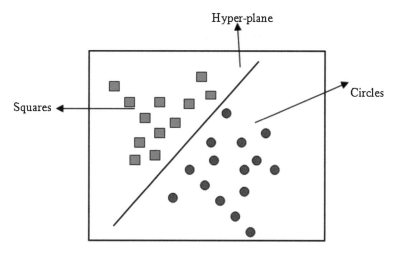

Fig. 9.14 Classification concept using SVM

The equation of hyperplane for classification of data using SVM is given by (9.12):

$$Y = W^{\mathrm{T}}\varnothing(x) + b \qquad (9.12)$$

where "W" is the normal vector of hyperplane and "b" is the offset vector. The SVM employs different kernel functions such as polynomial kernel, linear kernel, sigmoid kernel, and Gaussian kernel functions. Thus usage of SVM in image processing includes the following approach:

- Computation of features from known images.
- Train SVM.
- Obtain features from test image.
- Classify using trained SVM.

This work employed linear kernel function for classification. Initially we train the SVM using features obtained from 30% of the total document images of each language. The trained model is then used for testing the samples.

9.4   Experimental Results

To evaluate the proposed method, a data base of 1006 document images of Kannada, Marathi, Telugu, Hindi, and English are considered. The database is built by collecting document images from the textbook, newspapers, and Internet. These document images comprise of printed text, graphics, symbols with various sizes, and resolution.

Table 9.1 shows the details of the database. Documents belonging to each language are considered as different classes. The database has 197 Kannada, 184 Marathi, 198 Telugu, 216 Hindi, and 211 English document images. Figure 9.15 shows the sample document images of the database used for classification.

Table 9.1 Details of the database

Sl. No.	Language	Documents
1	Kannada	197
2	Marathi	184
3	Telugu	198
4	Hindi	216
5	English	211
Total number of document images		**1006**

(a)

Fig. 9.15 Sample document images from the database. (**a**) Kannada document images. (**b**) Telugu document images. (**c**) Marathi document images. (**d**) Hindi document images. (**e**) English document images

(b)

Fig. 9.15 (continued)

The SVM classifier needs to be trained with known set of features before it is used for classification. In the proposed algorithm, from each class 30% of documents are employed to train the SVM model and the remaining are used for testing. The detection rate is used as an evaluation parameter to compare classification performance of each class and it is by equation (9.13):

$$\text{Detection rate} = \frac{\text{Number of document correctly classified}}{\text{Total number of documents}} \tag{9.13}$$

We also used average detection rate to compare overall performance of the methods, which is given by equation (9.14). It is an average of detection rate obtained for all the classes of documents:

$$\text{Average detection rate} = \frac{1}{N} \sum \text{Detection rate} \tag{9.14}$$

(c)

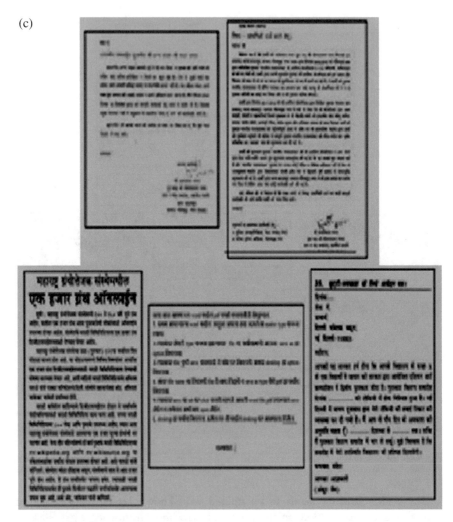

Fig. 9.15 (continued)

Where "N" is the total number of classes considered for a domain. In the presented work, as five classes are used, the value of "N" is taken as five. For testing, a document image from each class is given as an input and the result of the classifier is noted down. The results of the proposed technique are compared with three feature extraction schemes: rotation invariant LBP features [34], HOG features [28], and multi-resolution HOG features obtained using DWT and HOG. Table 9.2 shows the details of features used for classification with size of the feature vector.

Table 9.3 shows comparison of the results with k-NN classifier and Table 9.4 shows results obtained using SVM. From the tabulated results, it is clear that presented method provides better results with both the classifiers in comparison

(d)

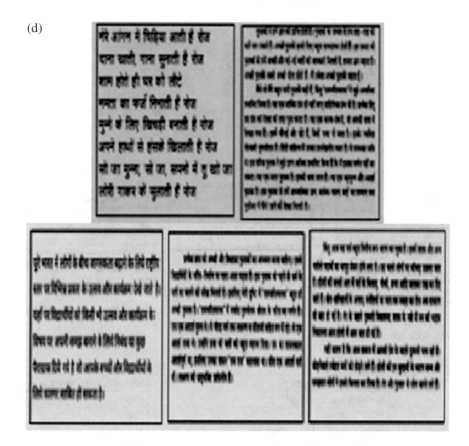

Fig. 9.15 (continued)

with existing methods. Particularly the proposed method with 128 × 128 cells provided better results with a feature vector of size 144, which is smaller feature vector compared to size of other feature vectors. It is found that the classification with SVM is much better compared with k-NN.

Figure 9.16 shows graphical comparison of the results with various feature extraction schemes using k-NN and SVM classifiers. The graphs are plotted for average detection rate versus feature extraction methods.

The observations revealed from comparison of the results are listed below:

- The proposed feature extraction scheme with cell size of 128 × 128 (144 features) performs better compared with existing state-of-art techniques.
- The SVM provides good classification accuracy compared with k-NN classifier irrespective of feature extraction schemes for this particular application.

(e)

Fig. 9.15 (continued)

Table 9.2 Details of feature extraction schemes

Sl. No.	Feature extraction method	Size of feature vector
1	Rotation invariant LBP	640
2	HOG features	324
3	Multi-resolution HOG features (DWT + HOG)	256
3	Proposed features with 64 × 64 cell size	1296
4	Proposed features with 128 × 128 cell size	144

Table 9.3 Comparison of results with k-NN classifier

Sl. No.	Language	Detection rate (%)				
		RLBP	HOG	Multi-resolution HOG features	Proposed features with 64 × 64 cell	Proposed features with 128 × 128 cell
1	Kannada	50	57.29	63.25	59.29	61.45
2	Marathi	56.31	77.49	81.43	68.93	81.55
3	Telugu	62.5	66.2	75.73	87.5	79.12
4	Hindi	98.4	94.11	97.97	99.49	100
5	English	84.82	83.33	77.78	82.2	88.48
Avg. detection rate		**70.406**	**75.684**	**79.232**	**79.482**	**82.12**

Table 9.4 Comparison of results with SVM classifier

Sl. No.	Language	Detection rate (%)				
		RLBP	HOG	Multi-resolution HOG features	Proposed method with 64*64 cell	Proposed method with 128*128 cell
1	Kannada	55.2	65.23	67.7	76.04	75
2	Marathi	62.135	73.28	73.78	85.44	87.38
3	Telugu	72.22	86.11	91.66	80.56	80.58
4	Hindi	98.99	95.23	97.46	100	100
5	English	86.44	79.56	80.126	81.15	92.15
Avg. detection rate		**74.997**	**79.882**	**82.1452**	**84.638**	**87.022**

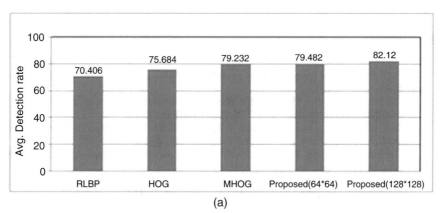

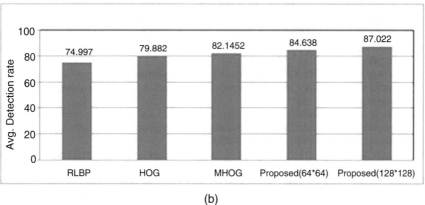

Fig. 9.16 Graphical comparison of results

9.5 Conclusion

This work proposed an efficient method for language-based classification of document images using SWT- and HOG-based hybrid texture features. It employs segmentation-free technique for recognition of documents' language. Proposed features are tested using K-NN and SVM classifiers on a database of 1006 document images. These features with SVM classifier provided an average detection rate of 87.02% for five different classes of document images comprising of Kannada, Marathi, Telugu, Hindi, and English language. Proposed feature extraction scheme can be tested on different scripts and also can be used in classification of other images with suitable preprocessing techniques.

References

1. G. Nagy, Twenty years of document image analysis in PAMI. IEEE Trans. Pattern Anal. Mach. Intell. **22**(1), 38–62 (2000)
2. S. Chaudhury, G. Harit, S. Madnani, R.B. Shet, Identification of scripts of Indian languages by combining trainable classifiers, in *ICVGIP 2000* (2000), pp. 20–22
3. A. Kulkarni, P. Upparamani, R. Kadkol, P. Tergundi, Script identification from multilingual text documents. Int. J. Adv. Res. Comput. Commun. Eng. **4**(6), 15–19 (2015)
4. M.C. Padma, P.A. Vijaya, Script identification from trilingual documents using profile based features. Int. J. Comput. Sci. Appl. **7**(4), 16–33 (2010)
5. U. Pal, B.B. Chaudhuri, Automatic identification of english, chinese, arabic, devnagari and bangla script line, in *Proceedings of Sixth International Conference on Document Analysis and Recognition* (2001), pp. 790–794
6. G.G. Rajput, H.B. Anita, Handwritten script recognition using DCT and wavelet features at block level. Int. J. Comput. Appl., Special issue on RTIPPR (3), 158–163 (2010)
7. S.M. Obaidullah, A. Mondal, N. Das, K. Roy, Script identification from printed Indian document images and performance evaluation using different classifiers. Appl. Comput. Intell. Soft Comput. **2014**, 1–12 (2014)
8. M.S. Shirdhonkar, M.B. Kokare (2010). Discrimination between printed and handwritten text in documents, in *IJCA Special Issue on Recent Trends in Image Processing and Pattern Recognition*, pp. 131–134
9. R. Pardeshi, B.B. Chaudhuri, M. Hangarge, K.C. Santosh, Automatic handwritten Indian scripts identification, in *IEEE 14th International Conference on Frontiers in Handwriting Recognition* (September 2014), pp. 375–380
10. C.L. Tan, W. Huang, S.Y. Sung, Z. Yu, Y. Xu, Text retrieval from document images based on word shape analysis. Appl. Intell. **18**(3), 257–270 (2003)
11. A.S. Wanchoo, P. Yadav, A. Anuse, A survey on Devanagari character recognition for Indian postal system automation. Int. J. Appl. Eng. Res. **11**(6), 4529–4536 (2016)
12. P. Sahare, S.B. Dhok, Script identification algorithms: A survey. Int. J. Multimed. Inf. Retr. **6**(3), 211–232 (2017)
13. U.D. Dixit, M.S. Shirdhonkar, A survey on document image analysis and retrieval system. Int. J. Cybern. Informat. **4**(2), 259–270 (2015)
14. S.A.A.A. Arani, E. Kabir, R. Ebrahimpour, Handwritten Farsi word recognition using NN-based fusion of HMM classifiers with different types of features. Int. J. Image Graph. **19**(1), 1–21 (2019)

15. N. Bi, J. Chen, J. Tan, The handwritten Chinese character recognition uses convolutional neural networks with the GoogLeNet. *Intern. J. Pattern Recognit. Artif. Intell.* **33**(11), 1–12 (2019)
16. C. Djeddi, I. Siddiqi, L. Souici-Meslati, A. Ennaji, Text-independent writer recognition using multi-script handwritten texts. Pattern Recognit. Lett. **34**(10), 1196–1202 (2013)
17. U.D. Dixit, M.S. Shirdhonkar, Fingerprint-based document image retrieval. Int. J. Image Graph. **19**(2), 1–17 (2019)
18. P.P. Roy, A.K. Bhunia, A. Das, P. Dey, U. Pal, HMM-based Indic handwritten word recognition using zone segmentation. Pattern Recognit. **60**, 1057–1075 (2016)
19. U.D. Dixit, M.S. Shirdhonkar, Preprocessing framework for document image analysis. Int. J. Adv. Netw. Appl. **10**(4), 3911–3918 (2019)
20. A. Bultheel, Learning to swim in a sea of wavelets. Bull. Belg. Math. Soc. Simon Stevin **2**(1), 1–45 (1995)
21. S.G. Chang, B. Yu, M. Vetterli, Adaptive wavelet thresholding for image denoising and compression. IEEE Trans. Image Process. **9**(9), 1532–1546 (2000)
22. A.N. Akansu, Y. Liu, On-signal decomposition techniques. Opt. Eng. **30**(7), 912–921 (1991)
23. M.J. Shensa, The discrete wavelet transform: wedding the a trous and Mallat algorithms. IEEE Trans. Signal Process. **40**(10), 2464–2482 (1992)
24. M.V. Tazebay, A.N. Akansu, Progressive optimality in hierarchical filter banks, in *Proceedings of 1st International Conference on Image Processing,* vol. 1 (Nov 1994), pp. 825–829
25. M.V. Tazebay, A.N. Akansu, Adaptive subband transforms in time-frequency excisers for DSSS communications systems. IEEE Trans. Signal Process. **43**(11), 2776–2782 (1995)
26. M. Holschneider, R. Kronland-Martinet, J. Morlet, P. Tchamitchian, A real-time algorithm for signal analysis with the help of the wavelet transform, in *Wavelets*, (Springer, Berlin, Heidelberg, 1990), pp. 286–297
27. Y. Zhang, S. Wang, Y. Huo, L. Wu, A. Liu, Feature extraction of brain MRI by stationary wavelet transform and its applications. J. Biol. Syst. **18**, 115–132 (2010)
28. N. Dalal, B. Triggs, Histograms of oriented gradients for human detection, in *International Conference on Computer Vision & Pattern Recognition (CVPR'05)*, vol. 1 (2005), pp. 886–893
29. Y. Zhao, Z. Song, X. Wu, Hand detection using multi-resolution HOG features, in *2012 IEEE International Conference on Robotics and Biomimetics (ROBIO)* (Dec 2012), pp. 1715–1720
30. Y. Zhao, Y. Zhang, R. Cheng, D. Wei, G. Li, An enhanced histogram of oriented gradients for pedestrian detection. IEEE Intell. Transp. Syst. Mag. **7**(3), 29–38 (2015)
31. X.Y. Li, Z.X. Lin, Face recognition based on HOG and fast PCA algorithm, in *The Euro-China Conference on Intelligent Data Analysis and Applications*, (Springer, Spain, 2017), pp. 10–21
32. J. Pan, Y. Zhuang, S. Fong, The impact of data normalization on stock market prediction: using SVM and technical indicators, in *International Conference on Soft Computing in Data Science*, (Springer, Malaysia, 2016), pp. 72–88
33. V. Vapnik, *The Nature of Statistical Learning Theory* (Springer Science & Business Media, 2013)
34. Mäenpaa Topi, Matti Pietikäinen. (2005) Texture analysis with local binary patterns, Handbook of Pattern Recognition and Computer Vision*, (pp. 197–216), Singapore: World Scientific Publishing

Chapter 10
Research Trends and Systematic Review of Plant Phenotyping

Bharati Patel and Aakanksha Sharaff

10.1 Introduction

Agriculture field is very popular for research work. It is essential in terms of food consumption for human needs and maintenance of the ecosystem for which it is required to give abundant attention towards the future challenges. Factors related to crop growth includes some of the major challenges are weather and atmospheric changes, which highly affect the agriculture field. Therefore planning should be in such a way that it assembles the precautions timely and required to develop a low-cost system model. Risks are not defined by the assumption-based approaches; therefore real-time data analysis is highly effective for all the concerns. Image processing is a technique which is based on real-time dataset analysis. The factors used for analyzing real time dataset are mainly time and cost, which are directly proportional to the efforts required for day-to-day analysis by manual observation. Now, this work is defining the challenge for huge dataset with less amount of effort – observation that is based on digitalized world and programming environment. Image processing is the area which has been implemented in plant phenotyping provocation very efficiently. Plant phenotyping is the method to extract the morphological characteristics or physical characteristics of the rice plant which are essential for yield production like height measurement and panicle counts. Rice plant contains tillers, leaf and panicle and panicle is the grain part which directly shows the counts of the panicles for one tiller of a single plant. Classification of inferior panicle count and superior panicle count exhibits an application of yield production and then regular automation is required to assess the quality of the grain part using plant phenotyping. Grain quality depends upon the stage of fertilization,

B. Patel (✉) · A. Sharaff
National Institute of Technology Raipur, Raipur, India
e-mail: bpatel.phd2018.cs@nitrr.ac.in; asharaff.cs@nitrr.ac.in

© Springer Nature Switzerland AG 2019
G.R. Sinha (ed.), *Advances in Biometrics*,
https://doi.org/10.1007/978-3-030-30436-2_10

which means if the seed is fertilized early, it means they are superior panicles and if later fertilized, they are called inferior panicles. Analysis of morphological characteristics is very important to improve the classification of best rice in terms of rice husk with its color and different shapes. Rice husk or the outer covering of the grain will show the grain quality as well as is useful on the power generation. Rice husk is difficult to use efficiently because of its properties: very hard surface covering, no food nutrition, and presence of very high silicon element; 20 percent of grain weight is attributed to rice husk, which is difficult to use as a biodegradable element [5]. It has been observed that different methods are following the same steps: first is to capture the image and then apply some programming techniques to measure the photographed scene, and after that, calculate the pixel values of the image in inches; afterwards, do the automatic scaling technique, which is setting an absolute scale. Therefore it is very crucial to analyze the rice growth according to its morphological properties. Many researchers have done the experiments based on plant height measurement, tiller counts, analysis of panicle architecture, and so on by using machine learning which is the most recent technique for direct automation. Image property can be enhanced by using the method of thresholding and filtering which are the main parts of the image preprocessing techniques. Regular automation of yield production requires very systematic approach to improve the accuracy in terms of timely observation of plant growth, early disease detection, etc. Error rate calculation of manual detection and direct automation will show the improvement of the technique. Most important factor to make the approach efficient when it should give the low error rate between the manual measurement and direct observation. Image processing is the best technique to take the image-based data for further observation. Image data means it should not be synthetic data. It should be real data for the observation so that it will give the information about the technique which produces less error rate method that is the primary concern because there are lots of methods which are available for digital image processing but unless it is useful for real time data then technique will not become effective. There are different ways to capture the digital images of rice crop; these are by using camera, computer scanning, satellite images, and drone-based images. After data collection, images are analyzed using a software that can easily calculate region of interest (ROI) and pixel value statistics based on user-defined parameters. Improvement of the image-based techniques will be shown by the fast analysis grading of quality of rice varieties which is a crucial point for market rate prediction. Automatic detection is the advantage of the digital-based image analysis because it is of low cost and less time observation. In this work most popular techniques are surveyed for the efficient analysis of the images such as deep learning, machine learning methods like random forest, support vector machines, convolutional neural networks (CNN), etc. Recently deep learning concepts for image processing are very effective for feature extraction; therefore this is a small effort to describe some methods of deep learning algorithms with greater efficiency which are related to rice crop features like panicle, leaf, tiller, root etc., are discussed in Table 10.1 and some selected algorithms are used those are not discussed for the analysis of the plant height, panicle counts and biomass calculation.

Table 10.1 Deep neural network model and its application – gap analysis

S. No.	Paper title	Gap analysis	Methodology used/Description
1	Plant leaf recognition using texture and shape features with neural classifiers [24]	Add some more combination of features other than shape, texture, and color. Using other classifiers like k-nearest neighbor, support vector machine, etc.	In this work shape- and texture-based analysis is computed by using Gabor filter and GLCM features. Curvelet features and invariant moments for shape analysis of the leaf
2	Smart farming: Pomegranate disease detection using image processing [25]	Some other combinations of feature like biomass, height, width with other deep neural network classifiers are required to estimate. Training time is slow for SVM	In this work preprocessing and feature extraction are done by using k-mean clustering and then using SVM classifier to classify the different classes of images
3	Plant species classification using deep convolutional neural network [26]	Size of training sample data is very low; therefore accuracy is not achieved. At the same time, SVM is slow for training the data. Feature like biomass is not considered	In this work deep convolutional neural network provides the pixel value calculation; feature extraction will be easy to segment by green pixel value rather than shape-based classification
4	How deep learning extracts and learns leaf features for plant classification [27]	Feature extracted based on shape, texture, color, and venation. But height, width, and biomass are not considered	In this work CNN is used for feature extraction, and deep net is used to species identification; therefore venation feature is effective to give species identification detail very accurately
5	A review of neural networks in plant disease detection using hyper spectral data [28]	It will give the information about the hyper spectral data, but features are required to match with the data. Therefore explanation of features is missing	In this work the architecture of all the neural network models is discussed with the purpose of disease identification. Learning vector quantization (LVQ) NN with PCA and RBF network with PCA are discussed for rice disease detection
6	Factors influencing the use of deep learning for plant disease recognition [29]	Factors which are influencing the performance are discussed here, but some more features are required to affect the method like biomass calculation and height of the plant	In this work method which is applied as transfers learning network which can reduce the numbers of the pre-trained network layers according to the effective result. Deep neural network will provide the effective result with all the considerations

(continued)

Table 10.1 (continued)

S. No.	Paper title	Gap analysis	Methodology used/Description
7	Method of plant leaf recognition based on improved deep convolutional neural network [30]	Complex background like biomass recognition is not identified	In this work image is preprocessed, and taking its effective size only so that the segmented image will give good result by using deep learning's layered approach
8	Tomato crop disease classification using pre-trained deep learning algorithm [31]	Features which are analyzed are size and weight; still some features like biomass will be effective for further analysis	In this work deep learning architecture will provide the stepwise analysis of tomato plant for disease detection. Transfer learning method is used for disease classification
9	High-throughput phenotyping with deep learning gives insight into the genetic architecture of flowering time in wheat [32]	Flowering time analysis is depend upon the regular automation of the plant, and it will be difficult for biomass, but it will be effective that at the same time bulk of analysis can be done by using biomass analysis	In this work CNN network is trained to analyze the image without wasting time to labeling the images
10	Three-channel convolutional neural networks for vegetable leaf disease recognition [33]	For disease detection, color, texture, and shape of the plant are used; at the same time, growth analysis is also one of the factors to analyze	In this work, each channel of TCCNN is fed by one of three color components of RGB diseased leaf image, the convolutional feature in each CNN is learned and transmitted to the next convolutional layer and pooling layer in turn, and then the features are analyzed through a fully connected network layer to get a deep-level disease recognition feature vector

First section is dedicated to introduction part and the second section, related work with the most recent work has done on image processing is described. In the third section, experimental setup has been discussed for automation of the crop field production. In the fourth section, results and discussion is described with a case study of highly efficient image processing techniques, and in the fifth section, conclusion and future work is discussed briefly.

10.2   Related Work

Plant phenotyping is a broad area which includes work on the plant varieties for improvement of the agriculture-based research field. Plant growth is correlated with the gene selection process for further improvement. Pasion et al. [1] proposed that computer-based technologies are showing the better result on processing plant physical properties like seed density analysis, panicle counts, spikelet's count, plant height measurement, grain quality assessment, gene binding assessment, tiller growth analysis, etc. Counce et al. [2] introduced grain quality assessment as major task to enhance the productivity of the rice crop. There are different techniques which are used to capture the different stages of the rice crop, and after achieving a particular growth, the seed fertilization is calculated in terms of seed density. For plant growth analysis it is necessary to recognize the stages, from stage 1 to stage 9, to know when the complete process of rice growth is done, but the main stage will come after stage 6; in this stage the rice grain is in mature condition. At stage 7 it has been observed that precaution is required because at this stage, plants are easily affected by the amount of water given, which will result in the damage of the grain and fungal infection and crucial disease infection. Identification of these problems has become a major issue to rectify the plant disease by using its properties e.g. white blast, brown spots etc. Singh et al. [3] discussed about various rice grain properties that will enhance the method of image based techniques to increase the productivity by regular assessment of crop field. Grain contains different properties, e.g., color, shapes, and quality; image analysis helps in predicting the good-quality grain production method. Atkinson et al. [4] proposed grain quality also depends upon the strong impact of internal root phenotyping of rice crop. Root phenotyping is the method to analyze the root of the crop according to climate changes as well as nutrition and water level assessment for better production. Previously it is implemented by the researchers that at stage 7, it is important to access the root functionality because in this stage, water level absorption can affect the grain quality; root phenotyping is one of the studious tasks. Zou and Yang [5] proposed that rice crop is very useful worldwide; therefore all the parts of the crop, from root to top, are point of the research. Rice grain is composed of rice husk, endosperm, bran, and germ. Rice husk contains high silicon, which is useful for power generation, but can be a source of environmental pollution, dust, smoke, and greenhouse effects. Panicle is another feature which is discussed by Zhou et al. [6] that image analysis techniques are very effective for tedious work like panicle architecture analysis of sorghum plant. Konovalov et al. [7] recommended image analysis techniques, which are effective in terms of less time automation with low cost. By using deep learning and machine learning concepts, it is easy to classify the problem and identify the precautions before the plant gets affected by diseases; automatic scaling of image will give the growth stage analysis of crop images. Cai et al. [8] introduced an approach to analyze the growth of the plant by using height calculation, which is the major task for growth analysis, and by using stereo image data, height is calculated very efficiently. Singh et al. [9] introduced deep learning concepts such as convolutional neural network which is used for the semantic segmentation for the images to preprocess the data with the effective result. Most popular image segmentation methods are like semantic segmentation, thresholding, region segmentation, edge based and clustering based segmentation and Ubbens et al. [10] discussed about image segmentation approach will provide the lack of complexity to analyze the area of interest so that by applying feature extraction method images will give the fast result analysis. Barbedo [11] proposed the image segmentation as the preliminary stage of image analysis which is very useful for noise minimization, and it will generate the low error rate also. Segmentation will give the focus upon the region of interest (ROI) which means only the selected and clear vision of the image. Jeon et al. [12] said the clarity of image is also one of the great issues recognized by the author to improve the image processing analysis. Jimenez-Berni et al. [13] processed information about plant biomass property, and it is easy for the ground surface because image clarity is not a big issue for the surface analysis. Most recent method for feature extraction is deep learning method and Kamilaris and Prenafeta-Boldú[14] also introduced about the deep learning concept for the complex feature extraction by using images. Malambo et al. [15] introduced an approach of density-based clustering; it is like biomass calculation of the images so it will show directly the yield production of the sorghum crop. Liakos et al. [16] introduced the complex feature calculation by machine learning techniques, which is also very useful for fieldwork analysis. Son et al. [17] introduced machine learning techniques such as support vector machine and random forest analysis for very effective analysis of yield production. Another useful technique for feature extraction is machine learning technique and Riegler-Nurscher et al. [18] introduced machine learning techniques which are useful for the pixel wise calculation, and for this technique image clarity is necessary. Bai et al. [19] introduced the multi-classifier cascaded methods, which are also used to analyze the image by using the SVM, gradient histogram, and CNN methods. In conclusion it has been observed by the study that plant growth analysis is based on its physical property analysis, and Tikapunya et al. [20] proposed grain physical characteristic measurement, and Santos et al. [21] proposed an approach to calculate rice grain dimension and chalkiness of the rice crop. Most of the papers are dedicated for height calculation and tiller count of the plant, which is very effective [22]. Sritarapipat et al. [23] proposed an approach for simple and effective baseline measurement for height calculation. Now, it is proven that many of the research works have done very effectively by their method of image analysis. Recently deep learning is very new and effective in terms of cost and time, and there are lists of papers which are based on classification and prediction-based model using deep learning. In Table 10.1, it is shown that some of the model is exclusively dedicated to limited numbers of parameters like color, texture, and size, but it will be effective if used in the growth analysis of the plant also. Some of the methods are direct approach toward the growth analysis of plants and also effective for the disease detection in terms of morphological property analysis. Table 10.1 shows gap analysis that focuses upon mainly three things according to future work crop analysis which are as follows:

1. Biomass analysis of rice crop by using deep learning algorithms
2. Plant growth analysis (plant height, panicle counts, tiller count) of rice crop correlation with grain density calculation by using deep learning algorithms
3. Comparison between feature extraction techniques and deep learning algorithms for huge amount of real dataset in relation to gene association of rice crop

In Table 10.1 some of the research papers have been discussed which is basically showing the disease-based method for classification and prediction instead of discussing about the features related to plant growth analysis. Therefore features are specific according to the analysis of the problem, and it is required to study the features which will give analysis of the plant growth as well as disease detection.

10.3   Experimental Setup

10.3.1   Dataset Gathering

(a) Dataset is taken from Indira Gandhi Krishi Vishwavidyalaya, Raipur.
(b) Drone-based images are captured with the height of 5 ft or 10 ft, but the clarity of the single plant is not accessible through the camera. Therefore it is the next task to make a clear vision of the drone-based images for field analysis.
(c) It is rice crop dataset with 30 images.
(d) Rice crop is planted in around 1 acre of 4046.86 square meter field area.
(e) Observation is started after 1 month of plantation, and images are taken within a week for the calculation of the plant height.
(f) Previous images are taken with the help of a drone-based system at the particular height of 5 ft or 10 ft from the ground in all the directions.
(g) Rice crop growth is hazy and complex and with uneven height creates the different image resolution sites.
(h) Drone-based images are not very useful to calculate the image height because the top view creates the problem.
(i) Therefore a single plant is taken for observation.
(j) For preliminary study, it is difficult to calculate the plant height for the whole field; the authors have chosen a particular area for capturing the images.
(k) Growth depends on the internal as well as external effects. Internal growth depends on the nutrients and water level basically and external on factors like climate and environmental changes.
(l) Plant growth is observed very slowly during the winter season, so panicles will generate after the month of March onward.
(m) Regulated observation by manual effort is time-consuming; therefore systematic image-regulated observation is required for the panicle counts.
(n) Panicle is the smallest unit of the rice crop, and at stage 7 it starts to get mature; therefore it is very important to take care of the rice crop after a particular growth in height.

(o) Every single plant has 4–5 tillers minimum, and each tiller has a number of spikelet which are the grain holder; therefore regulated observation is required by the researchers.

10.3.2 Preprocessing Module

Image processing has some predominant steps which are productive toward the image quality enhancement. Before using any technique, it is required to process the raw data by using preprocessing techniques:

(a) Preprocessing module uses the color thresholding to enhance the color quality of the image as shown in Fig. 10.3 for grayscale image.
(b) For image preprocessing, grayscale image has been taken for the further process instead of color images. Before going for height calculation, the image has some noises which are removed by the median filter with 0.5 errors, and then proceed to edge detection technique by using different techniques.
(c) Image is taken for background removal, so that pixel value can be calculated in an efficient way.
(d) Pixel value of the topmost point of the image will give the information about the height of the plant.
(e) In previous papers it has been discussed that plant height calculation is proportion to plant growth analysis [22] and height is calculated as the conversion of pixel value into inches.
(f) Preprocessing the images using the concept of deep learning includes change of size; direction of the image will be very easy to implement.
(g) Labeling of the image data by using train network for image classification is another approach to calculate its morphological properties.

10.3.3 Height Calculation Module

In this module preprocessed image is cropped for further calculation. Image is converted in grayscale, and then pixel value of the area is calculated. When method is applied for the cropped image, the selected area is masked, and then it is separately analyzed. Single plant height calculation is easy, and preprocessing will also take very less time. In conclusion, cropped area will give fast result as compared to the full image. Figure shows the calculated value of the pixel for the particular image.

10.3.4 Region of Interest Calculation

Region of interest will give the information about the particular plant top area so color image is converted into grayscale intensity image, and background is removed to capture the desired area. The desired area shows the actual region of interest (ROI). It will help to calculate the plant area as limited area will give efficient result to calculate the pixel value. For the field area, it will be difficult to calculate the separate region of interest (ROI) because of complex and hazy structure of the image. Drone-based image or the 3D image property calculation is required for the particular area of interest. Plant phenotyping is the application which is correlated with the gene association also; therefore classification of the image should be accurate to justify the physical property of the plant growth.

10.4 Results and Discussion

In this work it is described that the rice crop dataset has been taken from the Indira Gandhi Krishi Vishwavidyalaya, Raipur, and all the modules are described here. Result for single plant analysis is also showed. Rice crop images are taken with the particular timestamp so that growth can be easy to measure. For the single plant, drone images are not fully suitable for the experiment. Therefore front view images are taken for the experimental analysis of the rice crop images.

Figure 10.1 shows the image quality is improved by applying the method of median filter to remove the noise from the image. By applying this method into the gray intensity image, the color quality is improved. Image preprocessing is the part where the quality of the image is improved, and there are numerous stages for this method. Some other methods like erosion and dilation are used to improve the blur image. Figure 10.2 shows the edge detection method of the image by applying the different methods to select the best part of the image for the further analysis. Different methods are applied such as Roberts edge detection method, Sobel edge detection method, Prewitt edge detection method, Canny edge detection method, and LoG edge detection method. Above all the techniques, it has been

Fig. 10.1 Image filtering by using median filter

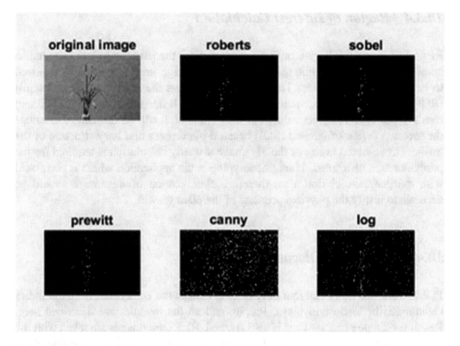

Fig. 10.2 Edge detection by using different techniques

concluded that Roberts, Sobel, and Prewitt methods are effective for edge detection as compared to other methods. It also shows the area which is highly demandable for the calculation of the pixels. Image contains pixel values in matrix formate which shows the range of the colors in the matrix form; so that it is observed that area which is detected having the different pixel values therefore it is important to take the right pixel value for the image. Sometimes real-time images are affected by environmental changes, and then the image quality will differ according to external factors like resolution of the camera, climate effect, different plant positions and directions, light reflection, and so on.

Real-time data testing is very useful for the new dataset because all the concerns are already tested by the researcher and in Fig. 10.3, it is tested means thresholding is the method of color balancing of the image. Image shows the effective approach to detect the region of the desired area. In plant image it is complex to recognize the leaf count and other panicle counts very easily. Edge detection technique will provide the density-based area to classify the image property. Feature extraction for a hazy field is not an easy task unless it has a better quality. Therefore image segmentation and background removal methods are useful if the image is cropped from the original.

In Fig. 10.4, the background is removed by the selection of the particular area. Then the grayscale image is cropped from the original image, and the selected area is now ready to give the details about the preprocessed image. In Fig. 10.5 it is

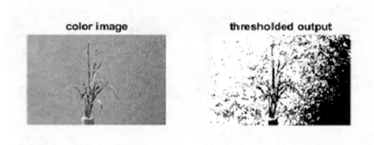

Fig. 10.3 Color thresholding for grayscale image

Fig. 10.4 Image masking technique

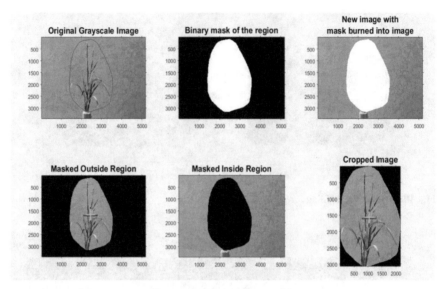

Fig. 10.5 Height calculation with pixel values and ROI of the image

Fig. 10.6 Pixel value
calculated values

Mean value within drawn area = 146.468
Number of pixels = 4502697
Area in pixels = 4503081.00
perimeter = 8529.91
Centroid at (x,y) = (2389.8, 1845.7)
Center of Mass at (x,y) = (2387.6, 1815.7)
Red crosshairs at centroid.
Green crosshairs at center of mass.

OK

shown that the image is masked, and the masked image is showing the segmented area of the original image. In Fig. 10.6 calculated values are expressed, and those values are as follows: the mean value of the selected area, number of pixels of the image, area in pixels calculation, total perimeter, centroid at point of (x, y), center of mass at point of (x, y), red cross marking at the point of centroid, and green cross marking at the point of center of mass. Therefore all the pixel value calculation is done; now the selected area removed by the method of centroid is separated from the original image. Original image shows the gray level image with the different pixel values, and according to that, the selected area is calculated. Figure 10.6 shows another method of image pixel calculation, so that manually it can be converted into centimeters or inches. Plant height calculation is showing the regular assessment

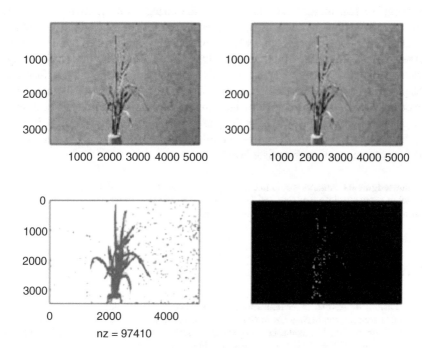

Fig. 10.7 Edge detection (Skeletonization) approach to calculate the pixel value for height calculation

of the plant growth, so that maturity level can be checked by the researcher to improve the plant growth. Necessary precautions are taken to capture the images like (early morning means around 8–9 o'clock) it is recommended to take the images at daytime, and after that, select 10 images for the testing and the 20 remaining images for training set.

In second method for image analysis is applied in Fig. 10.7 and that is to take the highest pixel value for the captured image, and same for the bottom level pixel value for the same image.

10.5   Conclusion and Future Work

This chapter mainly explores the important factors considered for yield prediction such as grain quality assessment and gene association analysis. It is dependent on various environmental factors like climate, weather analysis and internal plant growth analysis, i.e., root phenotyping, physical characteristic analysis, etc. Therefore the fastest approaches are required to capture the changes according to time and with regular automation. Such systems are which are free from manual calculation to design so that images can only capture the changes on the plants. Therefore programming techniques with imaging tools are going to be used by the researchers to make the automation effective with respect to time and cost. According to the survey, deep learning has been found a highly demandable approach for image feature extraction with less noise and also for huge dataset. It outperforms as compared to other techniques so it is highly useful for many of the areas like water level assessment, resource of nutrition scaling and for automatic scaling of the growth of the plant as well as grain quality assessment of the rice crop. But it is a great challenge for the researchers to implement the method on huge dataset with the fewer amounts of intermediate layers.

Acknowledgments Authors would like to thank the National Institute of Technology, Raipur, for providing necessary infrastructure and facility for the research. Special thanks to Indira Gandhi Krishi Vishwavidyalaya, Raipur, for helping us in this research work.

References

1. E. Pasion, R. Aguila, N. Sreenivasulu, R. Anacleto, Novel imaging techniques to analyze panicle architecture, in *Rice Grain Quality*, (Humana Press, New York, 2019), pp. 75–88
2. P.A. Counce, K.A. Moldenhauer, Morphology of rice seed development and its influence on grain quality, in *Rice Grain Quality*, (Humana Press, New York, 2019), pp. 57–74
3. A. Singh, B. Ganapathysubramanian, A.K. Singh, S. Sarkar, Machine learning for high-throughput stress phenotyping in plants. Trends Plant Sci. **21**(2), 110–124 (2016)
4. J.A. Atkinson, M.P. Pound, M.J. Bennett, D.M. Wells, Uncovering the hidden half of plants using new advances in root phenotyping. Curr. Opin. Biotechnol. **55**, 1–8 (2019)
5. Y. Zou, T. Yang, Rice husk, Rice husk ash and their applications, in *Rice Bran and Rice Bran Oil*, (Academic Press and AOCS Press, Urbana, Illinois, 2019), pp. 207–246
6. Y. Zhou, S. Srinivasan, S.V. Mirnezami, A. Kusmec, Q. Fu, L. Attigala, P.S. Schnable, Semiautomated feature extraction from RGB images for Sorghum panicle architecture GWAS. Plant Physiol. **179**(1), 24–37 (2019)
7. D.A. Konovalov, J.A. Domingos, R.D. White, D.R. Jerry, Automatic scaling of fish images, in *Proceedings of the 2nd International Conference on Advances in Image Processing*, (ACM, 2018, June), pp. 48–53
8. J. Cai, P. Kumar, J. Chopin, S.J. Miklavcic, Land-based crop phenotyping by image analysis: Accurate estimation of canopy height distributions using stereo images. PloS one **13**(5), e0196671 (2018)
9. A.K. Singh, B. Ganapathysubramanian, S. Sarkar, A. Singh, Deep learning for plant stress phenotyping: Trends and future perspectives. Trends Plant Sci **23**, 883 (2018)
10. J. Ubbens, M. Cieslak, P. Prusinkiewicz, I. Stavness, The use of plant models in deep learning: An application to leaf counting in rosette plants. Plant Methods **14**(1), 6 (2018)
11. J.G.A. Barbedo, A review on the automatic segmentation and classification of agricultural areas in remotely sensed images
12. G. Jeon, M. Anisetti, E. Damiani, O. Monga, Real-time image processing systems using fuzzy and rough sets techniques (2018)
13. J.A. Jimenez-Berni, D.M. Deery, P. Rozas-Larraondo, A.T.G. Condon, G.J. Rebetzke, R.A. James, et al., High throughput determination of plant height, ground cover, and above-ground biomass in wheat with LiDAR. Front. Plant Sci. **9**, 237 (2018)

14. A. Kamilaris, F.X. Prenafeta-Boldú, Deep learning in agriculture: A survey. Comput. Electron. Agric. **147**, 70–90 (2018)
15. L. Malambo, S.C. Popescu, D.W. Horne, N.A. Pugh, W.L. Rooney, Automated detection and measurement of individual sorghum panicles using density-based clustering of terrestrial lidar data. ISPRS J. Photogramm. Remote Sens. **149**, 1–13 (2019)
16. K. Liakos, P. Busato, D. Moshou, S. Pearson, D. Bochtis, Machine learning in agriculture: A review. Sensors **18**(8), 2674 (2018)
17. N.T. Son, C.F. Chen, C.R. Chen, V.Q. Minh, Assessment of Sentinel-1A data for rice crop classification using random forests and support vector machines. Geocarto Int. **33**(6), 587–601 (2018)
18. P. Riegler-Nurscher, J. Prankl, T. Bauer, P. Strauss, H. Prankl, A machine learning approach for pixel wise classification of residue and vegetation cover under field conditions. Biosyst. Eng. **169**, 188–198 (2018)
19. X. Bai, Z. Cao, L. Zhao, J. Zhang, C. Lv, C. Li, J. Xie, Rice heading stage automatic observation by multi-classifier cascade based rice spike detection method. Agric. For. Meteorol. **259**, 260–270 (2018)
20. T. Tikapunya, G. Fox, A. Furtado, R. Henry, Grain physical characteristic of the Australian wild rices. Plant Genet. Resour. **15**(5), 409–420 (2017)
21. M.V. Santos, R.P.O. Cuevas, N. Sreenivasulu, L. Molina, Measurement of rice grain dimensions and chalkiness, and rice grain elongation using image analysis, in *Rice Grain Quality*, (Humana Press, New York, 2019), pp. 99–108
22. K.P. Constantino, E.J. Gonzales, L.M. Lazaro, E.C. Serrano, B.P. Samson, Plant height measurement and tiller segmentation of rice crops using image processing, in *Proceedings of the DLSU Research Congress*, vol. 3, (2015, March), pp. 1–6
23. T. Sritarapipat, P. Rakwatin, T. Kasetkasem, Automatic rice crop height measurement using a field server and digital image processing. Sensors **14**(1), 900–926 (2014)
24. J. Chaki, R. Parekh, S. Bhattacharya, Plant leaf recognition using texture and shape features with neural classifiers. Pattern Recogn. Lett. **58**, 61–68 (2015)
25. M. Bhange, H.A. Hingoliwala, Smart farming: Pomegranate disease detection using image processing. Procedia Comput Sci **58**, 280–288 (2015)
26. M. Dyrmann, H. Karstoft, H.S. Midtiby, Plant species classification using deep convolutional neural network. Biosyst. Eng. **151**, 72–80 (2016)
27. S.H. Lee, C.S. Chan, S.J. Mayo, P. Remagnino, How deep learning extracts and learns leaf features for plant classification. Pattern Recogn. **71**, 1–13 (2017)
28. K. Golhani, S.K. Balasundram, G. Vadamalai, B. Pradhan, A review of neural networks in plant disease detection using hyperspectral data. Inf Processing Agric. **5**(3), 354–371 (2018)
29. J.G. Barbedo, Factors influencing the use of deep learning for plant disease recognition. Biosyst. Eng. **172**, 84–91 (2018)
30. X. Zhu, M. Zhu, H. Ren, Method of plant leaf recognition based on improved deep convolutional neural network. Cogn. Syst. Res. **52**, 223–233 (2018)
31. A.K. Rangarajan, R. Purushothaman, A. Ramesh, Tomato crop disease classification using pre-trained deep learning algorithm. Procedia Comput. Sci. **133**, 1040–1047 (2018)
32. X. Wang, H. Xuan, B. Evers, S. Shrestha, R. Pless, J. Poland, High-throughput phenotyping with deep learning gives insight into the genetic architecture of flowering time in wheat. bioRxiv, 527911 (2019)
33. S. Zhang, W. Huang, C. Zhang, Three-channel convolutional neural networks for vegetable leaf disease recognition. Cogn. Syst. Res. **53**, 31–41 (2019)

Chapter 11
Case Studies on Biometric Application for Quality-of-Experience Evaluation in Communication

Tatsuya Yamazaki

11.1 Introduction

With the advent of the information age, new communication services of high diversity emerge continuously. The demand of people for communication services is growing higher. Traditionally, the service level has been determined based on QoS (Quality of Service). QoS is the ability to provide different priority to different applications, users, or data flows. In other words, it is expected to guarantee a certain level of performance to a data flow, for instance, a required bit rate, delay, delay variation, packet loss, or bit error rates may be guaranteed. QoS sometimes refers to the level of service quality [1]. With the spread of smartphones, QoS is crucial for real-time streaming multimedia applications in a user mobile environment, for example, multiplayer online games and video streaming. These applications often require fixed bit rate and are prone to be affected by delay. QoS is especially important in a situation where the network resource capacities are limited.

Recently, QoE (Quality of Experience) is focused as a subjective metric to reflect user experience for the real-time streaming multimedia applications [2–4]. Since QoS is a quality measure evaluated from the service provider's point of view, it cannot directly describe users' satisfaction with services from the user's point of view. Nowadays, the service providers realize that they make much account of QoE to keep the users for their services. Therefore, QoE has become one of the important topics as a mobile communication ultimate metrics in the academic field as well as in the business field. The future development trend of telecommunication industry will become the key for the survival and profitability of communication operators,

T. Yamazaki (✉)
Niigata University, Nishi-ku Niigata, Japan
e-mail: yamazaki@ie.niigata-u.ac.jp

© Springer Nature Switzerland AG 2019
G.R. Sinha (ed.), *Advances in Biometrics*,
https://doi.org/10.1007/978-3-030-30436-2_11

that is, to develop full telecommunication service, enhance customer perceptibility, retain users, expand the size of users, and constantly improve market share.

Historically, QoS has preceded QoE. Before advent of QoE, communication services were objectively evaluated by QoS parameters such as packet loss rates, delay, delay variation, or average throughput, which has little relationship with customer evaluation. QoS is related to the network or communication system and the media that delivers contents. Then, in 2007, QoE has been defined as a subjective measure from the user's perspective of the communication service quality [5]. Laghari and Connelly [6] also referred that QoE needs to capture people's aesthetic and hedonic needs.

QUALINET (European Network on Quality of Experience in Multimedia Systems and Services) actively studied QoE under the COST Action IC 1003. As their working definition in 2014, QoE was defined as the degree of delight or annoyance for a service from the user's viewpoint [7]. This working definition was successfully included in recommendation of the International Telecommunication Union (ITU) as ITU-T P.10 [8] in 2016. Meaning of the user's viewpoint can be considered as human expectations, feelings, perceptions, cognition, and satisfaction for the service provided [9]. In other words, QoE is a blueprint of all human subjective quality needs and experiences for his/her particular contextual usage of service [6]. Although QoE is a subjective measure, communication operators desire to measure it to improve their service from the viewpoint of their customer.

One big issue is how to capture subjective evaluation for the service from user's viewpoint. As described in [3], one of the generic methods for QoE evaluation is MOS (mean opinion score), which is the average of values determined by subjects as their individual opinion by use of the predefined scale [10]. Usually the subject assigns his/her opinion after a set of experimental trial, so that MOS cannot record time-varying evaluation of the subject during the experiment. Therefore, in recent years, researches and developments of QoE pay attention to estimating QoE from human biological signals such as skin conductance activity [11] or EEG (electroencephalogram) [12, 13], where EEG was used for evaluation of the video streaming services.

In this chapter, we focus on Internet service quality evaluation by biometrics. In particular, QoE for mobile game services is evaluated by use of EEG biometric signals. The mobile game is targeted because data traffic of smartphone is rapidly increasing [14] as well as revenues from mobile gaming market are very promising in future [15]. In addition, Japan had is the third largest gaming market in the world in 2018 next to the first United States and the second China. The Japanese gaming market had grown to $19.2 billion. Since then, as the mobile gaming companies are expanding in Japan, its gaming market is expected to swell more. Therefore, QoE evaluation is also needed for the mobile gaming market and its related companies.

The purpose of this study is to analyze the relationship between the quality of the communication service and the biometric signal when the communication state changes in the mobile environment. As the QoS parameter that effects the quality of the communication service, delay is selected. In such a varying environment, we measure EEG of game players and collect their QoE evaluation. Simultaneous measurement of the biological signal and QoE under a time-varying communication condition is a new challenge in this area. We also verify the relationship between the biological signal and QoE.

11.2 Experimental Setting

11.2.1 Game for Experiment

Since existing real mobile games are affected by unstable factors such as the number of the online players, the distance to the online game server, and induced delay, it is very difficult to set stable network conditions. Therefore, an original RPG (role-playing game) was developed under a pseudo-QoS-controllable communication environment in order to obtain the relationship between EEG and QoE accurately.

The game process shown in Fig. 11.1 is divided into four modes: the introduction mode, the practice mode, the experiment preparation mode, and the experiment mode. Each mode is shown in Figs. 11.2, 11.3, 11.4, and 11.5, respectively.

The experimental procedure is as follows. When a subject executes this application, the game introduction mode automatically starts. In the introduction mode, explanation of the operation method is shown on the display, and the subject can read the explanation until the subject understands how to operate the game. Upon understanding the operation method, the subject pushes the play button to transit to the practice mode. During the 3 min practice mode, the subject operates a character from the starting point to the end point to attack monsters along the way of the mode task map shown in Fig. 11.6. After completion of the practice mode, it automatically transits to the experiment preparation mode. In the experiment preparation mode, seven buttons with the number from 1 to 7, respectively, are prepared. Seven groups with different QoS parameters values are set corresponding to the seven buttons. By pushing one from the seven buttons, the subject transits to the experimental mode with the designated QoS parameters. Details of the different QoS parameter values will be introduced in the following section. During the 2 min experiment mode, the subject operates a character from the starting point to the end point to attack the monsters along the way of the task map shown in Fig. 11.7. After completion of the experiment mode, it automatically returns to the experiment preparation mode, and then the subject selects the next button. This cycle lasts until all seven tasks complete.

11.2.2 Experiment Preparation

The subject is asked to play the original game with the wireless EEG measurement apparatus Polymate Mini AP108 (hereinafter, referred to as the EEG apparatus). The subject uses a tablet PC, which has no influence on the EEG apparatus, to operate the game task.

Fig. 11.1 Game process

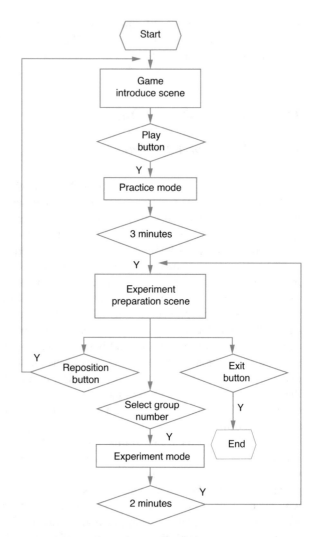

Fig. 11.2 Introduction mode

Fig. 11.3 Practice mode

Fig. 11.4 Preparation mode

Fig. 11.5 Experiment mode

Fig. 11.6 Practice task map

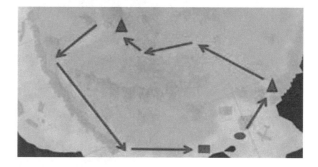

In the experiment setting shown in Fig. 11.8, PC(1) is used to control QoS parameters, PC(2) is used to record the EEG data, tablet PC is used to operate the game task, and the EEG apparatus is used to measure EEG. The experiment device information is shown in Table 11.1. An experimental scene is shown in Fig. 11.9.

Fig. 11.7 Experiment task map

Fig. 11.8 Experiment setting

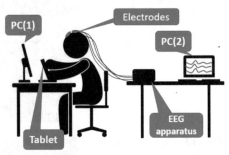

Table 11.1 Experiment device

Device	Model	OS
Note PC (1)	ASUS	Windows 7
Note PC (2)	Panasonic	Windows 10
Tablet PC	ASUS	Android 7.0
EEG apparatus	Polymate Mini AP108	

Fig. 11.9 Experiment scenes

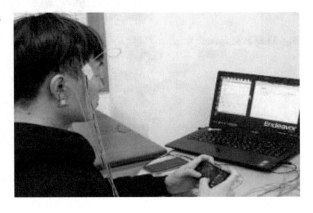

The numbers of subjects that have RPG experiences but are not experts are the following: 9 subjects of Chinese and 7 subjects of Japanese, totally 16 subjects. Each subject carries out seven game tasks.

11.2.3 Subjective Evaluation Experiment Procedure

Subjective experiment was conducted by SS (single stimulus) evaluation method [16], whose experimental process is shown in Fig. 11.10, where squares denote tasks, circles denote value sets of QoS parameters, and triangles denote subjective evaluation after completion of each task. The subject is asked to operate one task for 2 min, followed by a QoE evaluation period presented. Seven tasks are randomly presented to each subject.

Game task QoS parameter values are shown in Table 11.2. Task 1 is regarded as a benchmark without any deterioration. Task 2, Task 3, and Task 4 have a deterioration with delay; Task 5, Task 6, and Task 7 have a deterioration with character moving speed.

11.3 EEG Measurement

11.3.1 EEG Measurement Apparatus and Measurement Points

The wireless EEG measurement apparatus is shown in Fig. 11.11, where 10 small-scale ACT electrodes are available as the AP108 accessories. Six areas of the brain, Fp1, Fp2, F3, F4, FZ, and CZ, are selected as the measurement points of the EEG apparatus according to the international 10–20 electrode system shown in Fig. 11.12.

11.3.2 EEG Frequency Bands and Significance

EEG is a test that records the electrical activity of the brain. Normally, it is a noninvasive method placing the electrodes along the scalp. These electrodes pick up voltage fluctuations resulting from ionic current inside the brain [17].

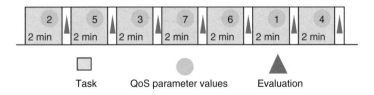

Fig. 11.10 Experiment scenes

Table 11.2 Task and QoS parameter values

	Task1	Task2	Task3	Task4	Task5	Task6	Task7
Delay (s)		0.0	0.6	0.9	1.5	0.0	0.0	0.0
Character moving speed (%)	100	100	100	100	80	60	40

Fig. 11.11 EEG apparatus

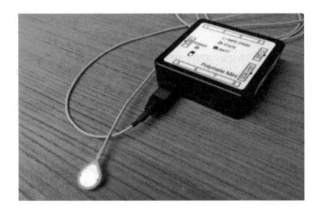

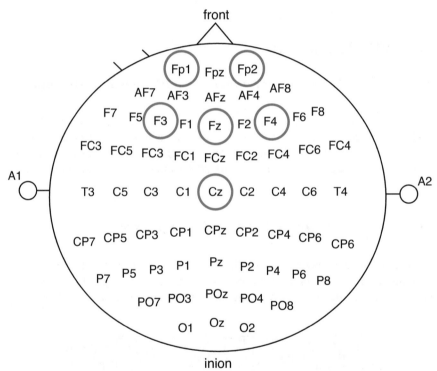

Fig. 11.12 Measurement points

The fluctuations are typically observed as waveforms of varying frequency and amplitude measured in voltage that is EEG.

EEG waveforms are generally divided into bands by frequency. Spectral methods are usually used to extract frequency bands, which are usually classified into bandwidths known as Delta (δ), Theta (θ), Alpha (α), and Beta (β) [18]. The EEG frequency bands are shown in Table 11.3.

Table 11.3 EEG frequency bands

Band	Frequency (Hz)
δ wave	<4
θ wave	$4 \leq$ and <8
α wave	$8 \leq$ and <14
β wave	$14 \leq$

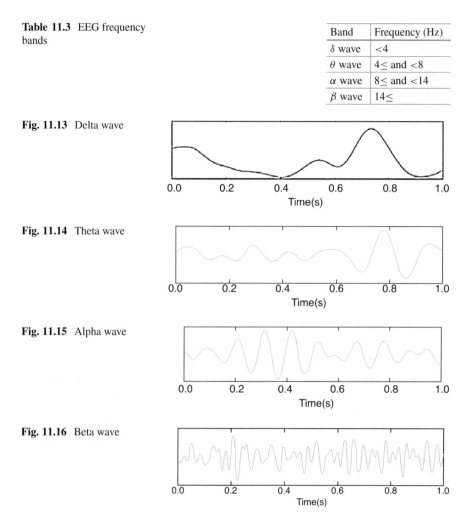

Fig. 11.13 Delta wave

Fig. 11.14 Theta wave

Fig. 11.15 Alpha wave

Fig. 11.16 Beta wave

The frequency range of δ wave is up to 4 Hz, shown in Fig. 11.13. It is the slowest wave, that is, its frequency is low, as well as its amplitude is the highest. Normally it is observed in deep sleep and meditation.

The frequency range of θ wave is from four to seven Hz, shown in Fig. 11.14. θ wave is observed during light sleep and meditation [19]. It is also observed in high concentration.

The frequency range of α is from 8 to 13 Hz, shown in Fig. 11.15. α wave is observed during meditation, relaxation, or contemplation.

The frequency range of β is from 14 Hz to about 30 Hz, shown in Fig. 11.16. β wave is observed during normal waking state of consciousness and is generally attenuated during active movements [20]. It can be divided into three sub-bands.

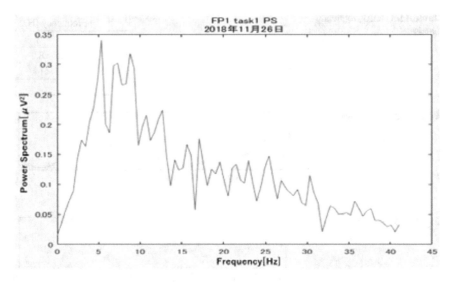

Fig. 11.17 FP1 position's EEG power spectrum for Task 1

11.3.3 EEG Power Spectrum

Through the experiments, we have obtained EEG time series data of 16 subjects.
After artifacts such as blinking are selected to cut out, a Butterworth filter of 4–30 Hz
is applied to five sections of the measured EEG waveform data. Consequently, each
section has 1024 points. Then fast Fourier transform is applied to the waveform,
and respective power spectrum is calculated. Finally, the power spectrum of the five
sections is averaged to calculate the final power spectrum results. For example, one
of the subject in every task FP1 position's EEG power spectrum is shown in the
following figures (Figs. 11.17, 11.18, 11.19, 11.20, 11.21, 11.22, and 11.23).

11.4 Experimental Results

11.4.1 EEG Power Spectrum Comparison

The second subject's FP1 position power spectrum for Task 1 and Task 3 are shown
in Fig. 11.24 as an example of the EEG analysis. In Fig. 11.24, the solid line presents
Task 1, and the dotted line does Task 3. Regarding Task 1, the measured main waves
are α and θ waves at FP1 position for all of 16 subjects. While α wave is the main
brain waveform in the state of awakening and relaxation, θ wave is the main brain
waveform in the state of high concentration. Both waveforms coexist in the situation
where the subject operates the game under normal network conditions. However,
when a QoS parameter value deteriorates, θ wave increases at FP1 position through
all experiments with a varying degree. We consider the reason for this phenomenon

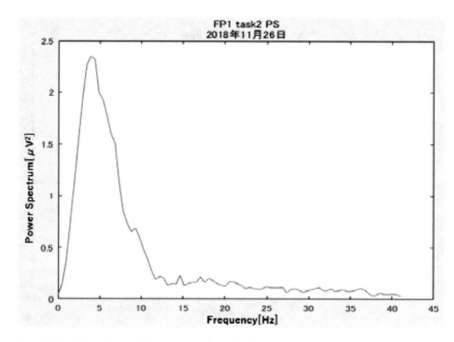

Fig. 11.18 FP1 position's EEG power spectrum for Task 2

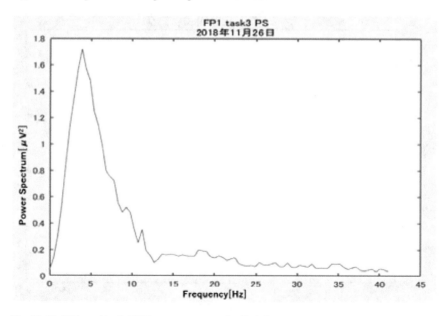

Fig. 11.19 FP1 position's EEG power spectrum for Task 3

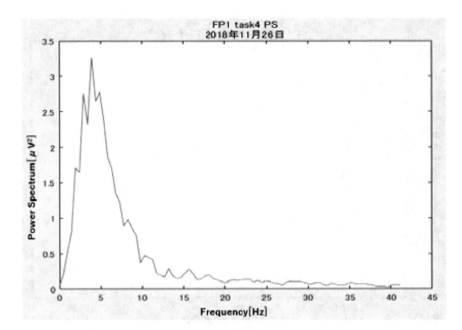

Fig. 11.20 FP1 position's EEG power spectrum for Task 4

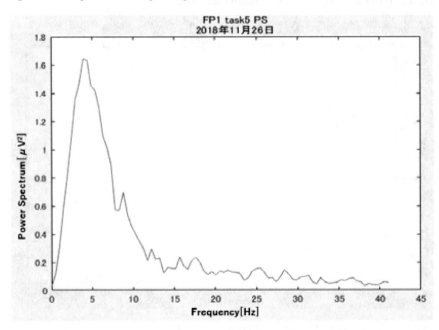

Fig. 11.21 FP1 position's EEG power spectrum for Task 5

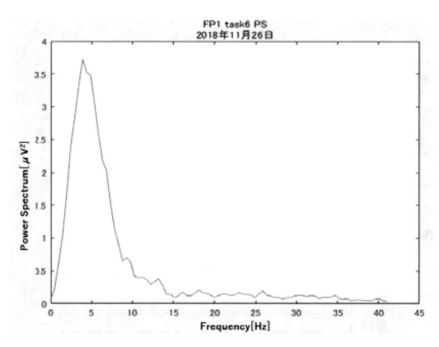

Fig. 11.22 FP1 position's EEG power spectrum for Task 6

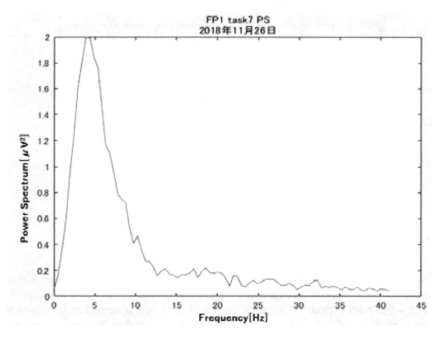

Fig. 11.23 FP1 position's EEG power spectrum for Task 7

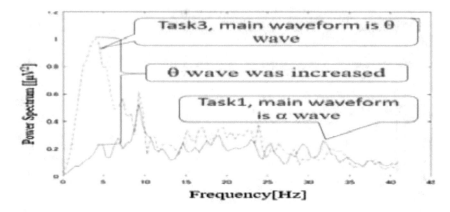

Fig. 11.24 Example of FP1 Task 1 and Task 3 power spectrum

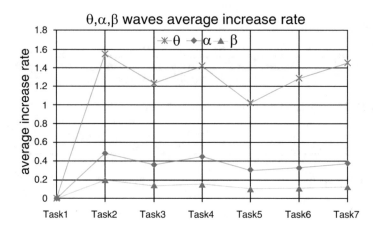

Fig. 11.25 Average increase rate

is that the subject must concentrate to accomplish the game task when he/she feels difficulty to play the game because of the QoS deterioration.

11.4.2 Average Increase Rate of Each Waveform

Regarding Task 1 as a benchmark, we calculated the 16 subjects average power spectrum value increase rate ([(the power spectrum value of each task in a specific band – the power spectrum value of specific Task 1 in a specific band)/the power spectrum value of Task 1 in a specific band]) of each task with three waveforms θ, α, and β, which are shown in Fig. 11.25. It is found that the θ wave has a great impact on the value of the QoS parameter which deteriorates than the other waves.

Compared with Task 1, the power spectrum value on the θ wave increases more than double. On the other hand, the average power spectrum values of α and β waveforms just a little increase. It can be observed that, in the process of mobile game operation, θ wave can be used as an important reference index to judge whether the QoS parameter values have deteriorated or not. It can also provide an important basis for predicting gaming QoE from EEG.

11.4.3 Average Increase Rate of Each Electrode Positions

We calculate the average power spectrum value increase rate of θ wave relative to Task 1 for each task at each electrode positions of 16 subjects, which is shown in Fig. 11.26. Obviously, when the value of the QoS parameter deteriorates, the θ waves at FP1 and FP2 positions have a very significant impact; both have increased more than twice in comparison with the average power spectrum value at other electrodes. Therefore, FP1 and FP2 can be regarded as the important positions of brain wave signal detection to help the prediction of gaming QoE. In the next section, we focus on the analysis of FP1 position and θ wave.

11.4.4 Comprehensive Comparison

In the experiments, two kinds of QoE were evaluated: a comprehensive QoE and an immersion QoE. In addition, the game score was recorded for each game task. These data are averaged for 16 subjects as comprehensive comparison. The

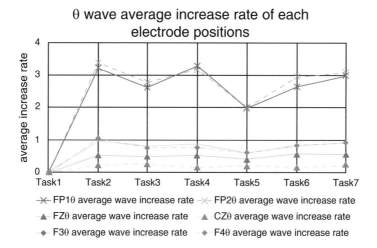

Fig. 11.26 Average increase rate of each electrode positions

Fig. 11.27 Comprehensive comparison

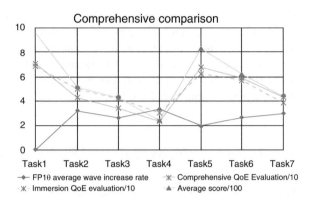

averaged results are shown with the increasing rates of θ wave at the position FP1 in Fig. 11.27.

It can be found that the QoS parameter values and the QoE values have proportional relationship. Namely, when a QoS parameter value decreases, the QoE values and scores also decrease. On the other hand, besides Task 3, the average power spectrum value of θ wave is inversely proportional to the value of the QoS parameter, when the QoS parameter value decreases, the average power spectrum value of θ wave increases. In other words, when a QoS parameter deteriorates, the subjects' attention becomes more focused. In next section, we will explain the reason why Task 3 is different with other tasks in the following analysis.

11.4.5 Player Level Comparison

Sixteen subjects are grouped as high-level players (HP) and low-level players (LP) by the averaged game score based on a threshold.

Regarding the delay, shown in Fig. 11.28, we can find that the comprehensive QoE evaluation values obviously differ between HP and LP in Task 3. HP seems to think that Task 3 and Task 2 are similar, because they can perform these tasks smoothly compared to LP and they spend similar attention to operate Task 2 and Task 3, so the θ wave increase rates of HP have similar tendency in Tasks 2 and Task3. On the other hand, LP thinks that Task 3 is much more difficult than Task 2. They cannot finish the task smoothly, so they begin to lose some interest in the game. So, their attention drops when they feel it is difficult to finish the task, and θ wave also appears to drop dramatically. Therefore, Task 3 is different from the other tasks in comprehensive comparison.

In addition, from Fig. 11.29, it is found that QoE evaluation trend regarding the character moving speed changing is similar between HP and LP for Tasks 5, 6, and 7. On the other hand, θ wave changing, however, obviously differs in Tasks 6 and 7 between HP and LP. Slower character moving speed may have a greater impact on

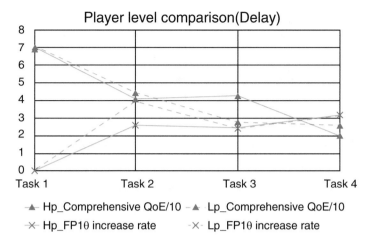

Fig. 11.28 Player level comparison of delay

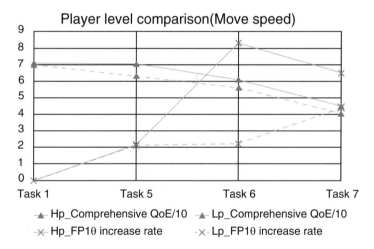

Fig. 11.29 Player level comparison of moving speed

HP. Although HP and LP have similar subjective perception for slow change of the character moving, HP will pay more attention to accomplishing the tasks even for the slow character moving.

Immersion QoE comparison between HP and LP is shown in Fig. 11.30. It is found that LP has tendency to show higher immersion than HP when the QoS parameter values deteriorate.

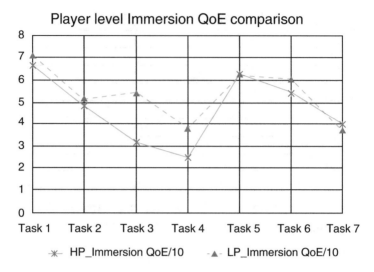

Fig. 11.30 Player level comparison for immersion QoE

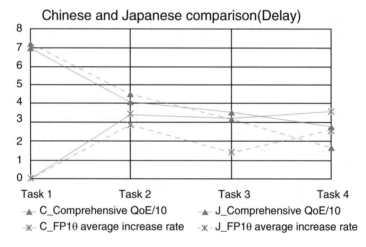

Fig. 11.31 Chinese and Japanese comparison of delay

11.4.6 Chinese and Japanese Comparison

Of the 16 subjects, 9 are Chinese, and 7 are Japanese. In this section we will analyze the differences between them.

Figure 11.31 shows that Japanese change in comprehensive QoE amplitude is more obvious with delay deterioration. This indicates that the Japanese are more sensitive to delay change when they operate an RPG. On other hand, Chinese power spectrum value increasing rate of θ wave is obviously higher than Japanese.

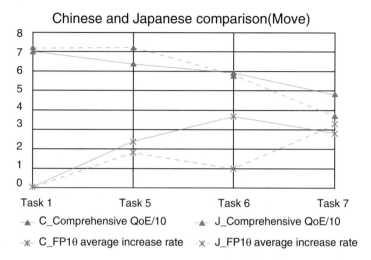

Fig. 11.32 Chinese and Japanese comparison of moving speed

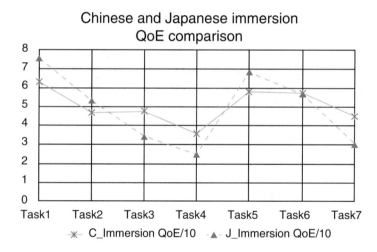

Fig. 11.33 Chinese and Japanese comparison of immersion QoE

In addition, regarding the character moving speed, shown in Fig. 11.32, the values of comprehensive QoE are almost the same as in the delay scenarios. The increasing rate of θ wave for the Chinese power spectrum value is also higher than Japanese obviously. And the immersion QoE comparison between Chinese and Japanese is shown in Fig. 11.33. Japanese change in immersion QoE amplitude is more obvious, so Japanese immersion QoE is more susceptible to the deterioration of QoS parameter values.

Because the number of participants in the experiment is not so large for each nationality, the results of the experiment cannot prove that there is a significant difference between the two nationalities. The significance is that individual differences in nationality, gender, and player level should be considered as a reference for QoE prediction from EEG.

11.5   Conclusions

Case studies of user-centric evaluation for a mobile game application have been conducted. EEG was used as a direct metric to capture user situation, and QoE evaluation was obtained from the user for different QoS situations. Through these experiments, we have verified that there is a certain relationship between EEG and QoE. It can be concluded that this relationship might be useful to predict user satisfaction through EEG. Of course, this prediction is based on only considering the impact of network traffic condition on user QoE while gaming but is excluding audio, game content, and other factors.

In a real situation, more rough classification of QoE can be useful as user satisfaction evaluation. For example, five scales of QoE (excellent, good, fair, poor, bad) might be enough to determine the user satisfaction level. On the other hand, individual differences must also have an impact on gaming QoE, so we should also consider the issue of individual differences in the prediction model.

This study has validated the relationship between QoE and EEG. Also, it can be found out what differences exist in the relationship between QoE and EEG when there are differences in the level of game players. The future ultimate purpose of this study is to predict the QoE of users through the relationship between QoE and EEG and apply it to the communication industry and the game industry.

Acknowledgements I gratefully acknowledge Mr. Qiu Han, who was a master's course student at Niigata University and contributed to this study so much. I also express my gratitude to Prof. Jun-ichi Hori in the faculty of engineering, Niigata University, for his professional technical support of EEG experiments.

References

1. M. Andreas, Real-time reconfiguration for guaranteeing QoS provisioning levels in Grid environments. Futur. Gener. Comput. Syst. **25**(7), 779–784 (2009)
2. Z. Duanmu, K. Zeng, K. Ma, A. Rehman, Z. Wang, A quality-of-experience index for streaming video. IEEE J. Sel. Top. Signal Process. **11**(1), 154–166 (2017). https://doi.org/10.1109/JSTSP.2016.2608329
3. T. Yamazaki, M. Eguchi, T. Miyoshi, K. Yamori, Quality of experience modeling with psychological effect for interactive web services, in *Proceedings of Second IFIP/IEEE International Workshop on Quality of Experience Centric Management*, pp. 1–4 (2014). https://doi.org/10.1109/ICME.2001.1237740

4. T. Hosfeld, R. Schatz, M. Varela, C. Timmerer, Challenges of QoE management for cloud applications. IEEE Commun. Mag. **50**(4), 28–36 (2012). https://doi.org/10.1109/MCOM.2012. 6178831
5. ITU-T: P.10/G.100 (2006) Amendment 1 (01/07): New Appendix I – Definition of Quality of Experience (QoE), *ITU-T Recommendation* (2007)
6. K.U. Rehman Laghari, K. Connelly, Toward total quality of experience: a QoE model in a communication ecosystem. IEEE Commun. Mag. **50**(4), 58–65 (2012)
7. P. Le Callet, S. Möller, A. Perkis, Qualinet white paper on definitions of quality of experience, in *European Network on Quality of Experience in Multimedia Systems and Services (COST Action IC 1003)*, Version 1.2 (2013)
8. ITU-T: P.10/G.100 (2006) Amendment 5 (07/16): new definitions for inclusion in recommendation ITU-T P.10/G.100, *ITU-T Recommendation* (2016)
9. K.U. Rehman Laghari, N. Crespi, B. Molina, C.E. Palau, QoE aware service delivery in distributed environment, in *2011 IEEE Workshops of International Conference on Advanced Information Networking and Applications*, pp. 837–842 (2011)
10. R.C. Streijl, S. Winkler, D.S. Hands, Mean Opinion Score (MOS) revisited: methods and applications, limitations and alternatives. Multimedia Systems **22**(2), 213–227 (2016). https://doi.org/10.1007/s00530-014-0446-1
11. T. Yamazaki, R. Yamamoto, T. Miyoshi, K. Yamori, Study of relationship between physiological index and quality of experience for video streaming service. IEICE Commun. Express **7**(6), 218–223 (2018)
12. S. Arndt, B. Kjell, E. Cheng, U. Engelke, S. Moller, J.N. Antons, Review on using physiology in quality of experience. Electron. Imaging **2016**(16) (2016). https://doi.org/10.2352/ISSN. 2470-1173.2016.16.HVEI-125
13. P. Seeling, Image quality in augmented binocular vision: QoE approximations with QoS and EEG. Period. Polytech. Electr. Eng. Comput. Sci. **61**(4), 327–336 (2017). https://doi.org/10. 3311/PPee.9454
14. Ericsson: mobility report. (Available via DIALOG, 2017). https://www.ericsson.com/en/ mobility-report/reports/november-2017. Cited 23 June 2019
15. Newzoo: Global Games Market Report. (Available via DIALOG, 2018). https://www. gamesindustry.biz/articles/2018-04-30-global-games-market-to-hit-usd137-9-billion-this-year-newzoo. Cited 23 June 2019
16. ITU-R Recommendation BT.500-13: methodology for the subjective assessment of the quality of television pictures (2012)
17. E. Niedermeyer, F.L. da Silva, *Electroencephalography: Basic Principles, Clinical Applications, and Related Fields* (Lippincott Williams & Wilkins, Philadelphia, 2005)
18. W.O. Tatum, Ellen R. grass lecture: extraordinary EEG. Neurodiagnostic J. **54**(1), 3–21 (2014)
19. B.R. Cahn, J. Polich, Meditation states and traits: EEG, ERP, and neuroimaging studies. Psychol. Bull. **132**(2), 180–211 (2006)
20. G. Pfurtscheller, F.H. Lopes da Silva, Event-related EEG/MEG synchronization and desynchronization: basic principles. Clin. Neurophysiol. **110**(11), 1842–1857 (1999)

Chapter 12
Nearest Neighbor Classification Approach for Bilingual Speaker and Gender Recognition

Samrudhi Mohdiwale and Tirath Prasad Sahu

12.1 Introduction

Speech is the primary form of human communication. From the evolution of telephone, speech is transformed into electric signals by the transducer in order to increase the reachability and enhance modes of communication [1]. Processing of audio signal is important to understand the path between the speaker and listener. The initial step of communication is a thought which is transformed into words, sentences, and phrases according to the grammar of the particular language. A thought comes according to the situation of the surrounding; this excites nerves of the brain to generate electrical signals which further excites the vocal chords and muscles of the vocal tract. This results in vibration as a change of pressure in the vocal tract and lip movement depends on that process of pressure change. Finally, lip movement transfers the generated thought in the form of speech over the space [2]. Space is a medium of communication and it has certain characteristics which are important specially in the area of forensics to recognize the background. Space contains different undesired sound that affects the intelligibility of speech. In order to improve speech intelligibility, noise should be removed from the recorded sequence. The enhancement of speech signal by noise reduction is very effective if enhancement approach is modified according to the type of noise [3].

In the era of handset, speech signal is transmitted in the form of electric signal. These are transformed and decoded by the transducer. Now in the digital era of communication analog to digital converter has introduced to digitally transmit and process the speech signals. Digital technology with its high speed and low cost with reduced power consumption replaces the huge part of analog-based technology. The

S. Mohdiwale (✉) · T. P. Sahu
National Institute of Technology, Raipur, Chhattisgarh, India
e-mail: tpsahu.it@nitrr.ac.in

© Springer Nature Switzerland AG 2019
G.R. Sinha (ed.), *Advances in Biometrics*,
https://doi.org/10.1007/978-3-030-30436-2_12

Fig. 12.1 Analog to digital conversion of speech signal

term digital speech signal processing defines the process of change of analog speech to digital speech that is having desired properties required for further applications. Analog to digital conversion of speech is shown in Fig. 12.1.

Speaker Recognition and Reason for Its Popularity

Speaker recognition thought is initiated from a clue on how a human can communicate with the device without any physical contact such as a keyboard or mouse. The machine can understand digital information and in order to personalize the machine, the speaker should recognize correctly. Everyone in the world wants a sophisticated life with a personalized robot who can work on their single command. Research is nowadays going toward that direction.

As we have seen speech is the most common and effective way of communication between human beings. A speaker plays a very important role in communication to transfer information. Identity of a speaker has a significant impact on research area, since it is a billion-dollar industry and has an excellent opportunity in various application fields. In the digital arena wide-range applications of speaker recognition include personal assistant in mobile and other devices, robotic control, security and forensics, health, and education [1–5]. Smart industries now work on classification of gender and age group of speakers to advertise accordingly for their product [4, 5]. Automatic speaker recognition system verifies the speaker identity with the samples of known identity. Speech is a nonstationary signal, hence hard to process; to work on speech signals small samples of few milliseconds are taken which make them stationary for a short duration of time. Precise speaker recognition focuses on feature extraction and classification methods to enhance the existing models. A variety of features are investigated to classify speakers, Few of them are Pitch intonation, lexical information, prosody which is termed as high-level feature. These are easy to extract from humans but complex to extract via machines. Acoustic spectral features, vocal track length and resonance, and glottal flow are low-level features easy to extract by short-time Fourier transform (STFT) or mel frequency cepstrum coefficient (MFCC), etc. [6]. In industrial use of speech signal processing audio indexing, baking authentication, information retrieval, remote monitoring, and forensics are some popular areas which explore audio processing [4, 5]. Even though it's very popular, it has many research challenges that excite researchers to explore the area.

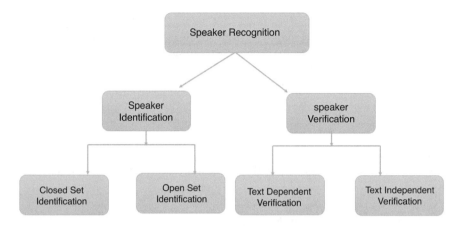

Fig. 12.2 Variants of speaker recognition

12.1.1 Variants of Speaker Recognition

Till now we studied about speaker recognition. Speaker recognition also known as speaker biometrics is a comprehensive term which is used for a procedure to identify a person, by recognizing him with his/her voice. It consists of a variety of branches which are either directly or indirectly related to each other. Speaker recognition variants are shown in Fig. 12.2.

Speaker identification and verification are two variants in which speaker recognition can be classified. The details for each are described below:

Speaker Identification
Speaker identification refers to the process of identifying the unknown speaker. In the process of identification speaker's voice is compared against the set of available speakers. Since for P available speaker it compares one voice against P voices, hence it is also represented as 1:P method [7].

For better understanding let's take an example of dataset that contains five voices: A, B, C, D, and E. If any one of them, suppose A, speaks the identification system, compare the voice of speaker A with all five voices. The identification model produces the output by comparing A's voice with available set of voices based on best similarity of voice.

The identification system of a speaker can be classified into two types:

1. Closed Set Identification

    In this type of identification, the speaker is one of the claimants from the group of members. No new member is introduced for identification. If the unknown speaker is not available in the database, the identification system still assumes the unknown voice is one of the speaker's voices from the available database of speech. This process consumes more time than speaker verification system because it is having multiple comparisons of voices in available database to identify the person. This type of identification is used to find the person that belongs to which group in the set of available groups. One disadvantage of this process is that unknown person cannot be identified [2].

2. Open Set Identification

In this type of identification system unknown voices can be introduced. If any unknown speaker speaks, the identification model shows the result "no match found" or "speaker is unknown." The total number of speakers is also unknown in this process [2].

Speaker Verification

Speaker verification refers to the method of verifying the identity of the speaker who is being spoken (test speaker) by nonspeech techniques known as content-dependent techniques. The voice of the speaker is matched with the similar voice present in the dataset; the similar voices are very less generally one or two, so the number of comparisons involved in verification task is less than identification task. This reduces the time complexity of the system [7]. Speaker verification can be done based on text-based features or voice-based features. So speaker verification can be classified into two types:

1. Text-Dependent Speaker Recognition

This type of speaker recognition requires the speaker to say the exactly given or commanded password. This type of speaker recognition has a robust approach. It only applies to the speaker verification branch. Most of the other branches are in a passive manner; therefore text-dependent speaker cannot be used in various branches. Another use of this type of speaker recognition is that the liveness problem of text-dependent system can be fixed by text-prompted modality [2, 7].

2. Text-Independent Speaker Recognition

It is one of the most versatile modalities and most feasible modalities which can be used in all branches of speaker recognition. This recognition system is independent of text being spoken by speaker and verify the identity of speaker by glottal features irrespective of text. It includes different categories into it. It also faces a few problems; one of the common problems faced by it is the possibility of a poor coverage of a part of speech. Another problem faced by it is the liveness assessment problem. It is also one of the common problems faced in text-independent system [2, 7].

12.2 Applications of Speech Processing and Speaker Recognition

Nowadays speech recognition (SR) technology industry has developed a broad range of commercial products in which speech processing technology is used efficiently and proves to be very useful. The historical uses of SR technology include dialing numbers by speech, call routing, response of a call with voice commands, speech to text conversion for fast data entry, voice-controlled devices, games, voice search options, transcription, and robotics. These applications are limited for speech with limited training samples. As the demand of easy and handled resources increases big data came into picture. Computational power with GPU support is also available to handle big data. So by using these facilities personalized assistance on your mobile and PC is introduced; Google now, Siri, and Cortana are the recent developments in the area of SR. Home automation, robotics, and machine translation are growing area nowadays. Few applications are described in detail below.

1. Security

   Security of belongings and property is very much essential. Many companies and residents use different kinds of security systems which are UI-based ID/password for protection. But hacking of this ID/password is happening very frequently and this system is tagged as unsecure. This increases demand of such a system which will not depend on any ID/password. Speaker's voice comes out as the solution to this problem. In speaker recognition system identification of the speaker as well as its verification is based on biometric features of vocal system of humans which will be safer than previous security systems. On the other hand, speaker recognition-based systems are also helpful in criminal cases for investigation of fraud against the voice of the speaker [4]. For practical purpose the voice-based security systems should be built with care of acoustical mismatch at verification phase.

   If the case arises of twins, one study shows that features of twin voices are also unique, so this can be used in administration system [8]. One of the parametric representations of speech is mel frequency cepstral coefficient which gives better results in speaker verification systems for speaker recognition task [9].

2. Crime investigation

   As security issue is discussed above one of the major areas of application of speaker recognition system is for crime investigation. Although there are many tests such as Norco test available, speaker recognition will be very easy and less expensive. This could be possible against voice samples available at crime place or its mobile phone recording for investigation.

3. Personalized robots

   Various patients suffer from very serious disorders and are not able to work by their own. At this time personalized robots are very helpful which understand the direct command just by voice and work will be done with command only. Recently this feature is also available for everyone. Alexa is one of the best example of it.

4. Medical field

   In medical fields this technology is used for medical transcription, electronic medical record, and many more [10]. Medical science frequently deals with patients and nowadays distance medicine can also be supplied based on question answering (QA) with patients. This QA session in text is a very tedious task for doctors in their busy schedule so speech recognition systems will be helpful for them. At the same time speaker's voice is also helpful to detect many problems. So the speech processing is applied in this area.

5. Education

Speech recognition software is in demand in the market as it can transcript the speech into text. This is very helpful for the students who are not able to listen. The transcripted lectures can be displayed to them. These lectures also work as notes for the students to revise the concepts [11].

6. Smart homes

Today the homes are assumed to be smart enough to save energy and time. In smart homes all the devices are work with owner's command. This is all in one application of speaker recognition in which privacy as well as personalized robot is covered.

12.3 Limitations of Speaker Recognition

1. *Not applicable for inarticulate persons*

As voiceless persons are unable to speak, one cannot obtain speech from them; hence speaker recognition project is not applicable to those persons. This is one of the major drawbacks of speaker recognition devices.

2. *Health issues may create verification problem*

Speaker verification directly depends on speaker's voice so any health issue such as throat problem, cold, and cough may cause problem at the time of real-time speaker verification.

3. *Effect of aging on voice*

As the age of person increases from childhood to adulthood, the voice along with its pitch and modulation also change. This has become a major challenge for voice biometric systems.

12.4 Issues and Challenges

- *Nonstationary Signal Processing*

  Acoustic signal changes its frequency with respect to time. The speech signal didn't have gaps between the spoken words; this make it difficult to determine the word boundaries. To solve the problem of nonstationary nature of acoustic signal processing, speech signal is divided into small time stamp (generally in ms) in which speech is assumed to be stationary [12].

- *Lack of Relevant Data*

  Voice biometrics is one of the latest fields of research but it's hard to find speech dataset of the same speaker from his childhood to old age. Similarly, in the field of linguistics relevant data for bilingual or multilingual speaker with proficiency in both the languages is not available, especially for native languages. This has become a challenge for speaker recognition.

- *Noise Removal*

  The speech itself is a very challenging domain due to its nonstationary nature but noise makes it more challenging. As noise constantly corrupts the speech signal and makes it difficult to process, an efficient method is important for noise removal to get better and optimal performance in speech processing task. Various methods have been proposed for noise removal [8, 9].

- *Multilingual Speaker Recognition*

  Text-independent speaker recognition model is considered as independent of language being spoken but its performance will be affected by multilingual trail conditions [15]. Since the large English corpora are available with multiple speaker recordings with different degradation conditions, therefore it shows better performance for English than any other languages. Lack of multilingual dataset restricts the model and is one of the causes of performance degradation [6].

- *Psychological State of Mind*

  Speech is directly affected by thoughts that came out of the mental state such as emotion, stress, anxiety, health, etc. These psychological states change the modulation of the speech signal and affect the feature of voice that characterizes different speakers [4, 16, 17].

Motivation

Speaker recognition system is a part of voice biometrics. In real-time scenario, this voice biometric system is not yet commercially implemented due to its reduced performance compared to other biometric systems. These are not yet fully reliable but the study reveals that speaker recognition-based systems are highly accessible and acceptable with ease of remote monitoring and low cost [13, 14]. To make this system reliable a perfect model is required which will provide effective preprocessing with faster response time. The system should conform to any situation. These requirements are the source of inspiration to provide better solutions for preprocessing techniques. India is a country of diversity. Many speakers are multilingual and most of the population is using a mix of two or more languages (e.g., Hinglish) instead of any single language, which increases the curiosity about the effect of language on automatic speaker recognition model. Different research on language-independent model encourages us to work in the area of speaker recognition.

Various classification methods such as support vector machine (SVM), Gaussian mixture model-universal background model (GMM-UBM), deep neural network (DNN), and deep belief networks (DBF) have been used for speaker classification [14, 15]. k-nearest neighbor (k-NN) classifier is also one of the effective classifiers with ease of implementation and good generalization ability. Ensemble methods are used to improve the instability of classifier as well as improve k-NN by subspace method by using sensitivity of input space. This improves overall performance of k-NN classification [4].

The rest of the paper is organized as follows. Section 12.2 represents related works, Sect. 12.3 describes proposed methodology for gender and speaker recognition, Sect. 12.4 represents results of classification, and Sect. 12.5 gives the conclusion of the presented method.

12.5 Related Work

Von Kempelen was the first one who demonstrated that human speech production system could be modeled. The first device based on human speech synthesis made in 18th century which is also called as speaking machine. The machine responds to all sounds and words other than Rex but it was found that this also responds to nonspeech signals which have 500 Hz energy. Rex was unable to reject the words which are not available in the vocabulary. Further in 1958 Duley created a classifier which works on spectra rather than formants. Grammar probabilities are added by Dene and from the recognition of few words or sounds the speech recognition concept evolved and enhanced in each decade [21]. Every individual has some specific vocal characteristics which may depend on linguistic means or utterances or may not. Vocal tract, articulator movement, gender, and pitch are some characteristics which allow the listener to identify the speaker. Researchers are more attracted toward speaker recognition due to its potential applications in the area of intelligence, fraud detection, authentication, and many more. Speaker and gender recognition is related area in our current work. Abbas Khosravani et al. [6] proposed multilingual speaker recognition with probabilistic linear discriminant analysis (PLDA) for text-independent speaker recognition category. The paper uses DNN-HMM approach for voice activity detection; i-vectors have been used as features with language-independent PLDA approach to minimize inconsistency in calculated features. Rosa Gonzalez Hautamaki et al. [22] discussed the effect of voice altered by age and its effect on speaker recognition; this shows fundamental frequency relatively less affected than other formant frequencies. While considering both male and female speakers fundamental frequency increases in both the cases but more increment shows in female voice during alteration. Vocal tract outline also has a significant impact of voice disguised in first four formant frequencies. Gang Liu et al. [23] proposed a novel approach to noisy and multi-session enrolled data with five back ends for speaker recognition system with extrinsic and intrinsic back end for extremely discriminative speaker recognition model. Authors also explore more comprehensive set of features for small dimensional i-vector models with diversify contents of information before modelling. Saeid Safavi et al. [24] work on children voice for classification of gender, age, and speaker identification. I-vector-based GMM-UBM and GMM-SVM models are compared for performance evaluation. Spectrum region with most significant information is investigated. This shows 0.9–2.6 kHz frequency band is for gender identification which is the second formant location while 1.6–3 kHz range is useful for age group identification. The human brain collectively interprets the speaker and speech while listening to any voice. To work simultaneously with these two Zhiyuan Tang et al. [25] proposed multitask recurrent model with collaborative joint training framework. To enhance performance of individual task simultaneously collaborative tasks interchange information accordingly. This method presents another neural network approach for classification. Eduardo Lleida et al. [26] discussed special issues in the area of speaker recognition. The paper focused on issues like spoofing, channel mismatch, spoofing countermeasures, and short speech utterances. Also discussed are state-of-art technologies in speech recognition field. Ankur Maurya et al. [27] present MFCC-GMM approach of speaker recognition with Hindi speech signal. It also discussed the challenges faced during recognition process including psychological and physiological challenges.

12.6 Proposed Method

Bilingual dataset is created to perform experiments on change in language of the same speaker. MFCC feature is extracted and classification is done using the methods described below. Detailed description of methodology is presented in subsequent sections.

12.6.1 Dataset

The dataset for bilingual speaker and gender recognition is created for the research. The dataset consists of 20 speakers; ten male and ten female speaker voices have taken for the research. Voice samples are taken via different mobile phones and in different environments to check the performance of the model in different real-time environments. Speakers are said to record samples in Hindi and English language with given text. The 2-D distribution of dataset is shown in Fig. 12.3. Each color represents MFCC distribution of different genders for gender recognition shown in Fig. 12.3a and MFCC distribution of 20 speakers for speaker recognition shown in Fig. 12.3b. The figure shows distribution of first two consecutive features for both gender and speaker recognition. For all 16 features this combination can be judged.

12.6.2 Flow Chart

The flowchart below shows the flow of proposed method for speaker and gender recognition task (Fig. 12.4).

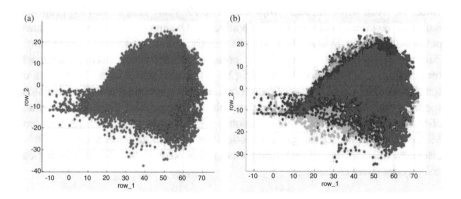

Fig. 12.3 Feature distribution of gender and speaker recognition, respectively. (**a**) Original dataset: gender classification, (**b**) Original dataset: speaker classification

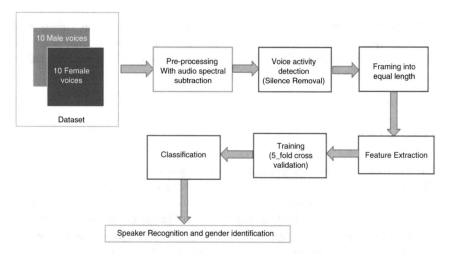

Fig. 12.4 Flowchart of proposed method

12.6.3 Methodology

In this section each phase of proposed method is discussed in detail. Starting from preprocessing, how it is performed for the given dataset, method for silence removal used, framing, feature extraction, and classification methods employed are presented below.

- *Preprocessing*

    The initial step of speech processing is the removal of noise from speech signal so that noise cannot degrade the recognition performance. For preprocessing of speech spectral subtraction method is employed. To perform spectral subtraction speech

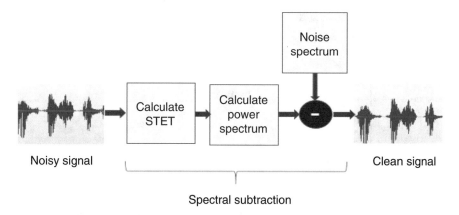

Fig. 12.5 Preprocessing using audio spectral subtraction

is assumed to be uncorrelated with additive noise and have zero mean [28]. The process of spectral subtraction is shown in Fig. 12.5. In this method, filtering of signal gives the less attenuation to high SNR than low SNR of speech sample.

As noise is assumed to be additive for spectral subtraction method, noise spectrum can be subtracted from noisy speech to get the clean speech. Noisy signal is first divided into frames of small window length typically 400 samples with different windowing techniques such as hamming window, hanning window, etc., to make the signal stationary within short time stamp. Short-time Fourier transform or discrete Fourier transform is calculated on each frame further in the process of spectral subtraction. Magnitude and phase spectrum are obtained from these transforms; magnitude transform is used to estimate the power spectrum of the signal. It is assumed that few initial frames are silence frames where no speech is observed; these frames are used for noise spectrum calculation. Now we have both speech and noise spectrums which can be subtracted and clean speech has obtained [33].

- *Silence Removal*

Silence in the speech gives many unnecessary observations which lead to misclassification of data points and affect the performance of the system. Voice activity detection using zero crossing rate is the popular method for silence removal.

Zero crossing rate is the rate of change of sign of the speech signal or rate of crossing the time axis (0–midpoint). Noise has property of maximum zero crossing rate but as it is removed in the previous step only silence points exist in the signal which follows the time axis with value zero. In the silence time stamp zero crossing rate becomes zero and this leads to silence removal from the given dataset (Fig. 12.6).

Silence area   Silence removed

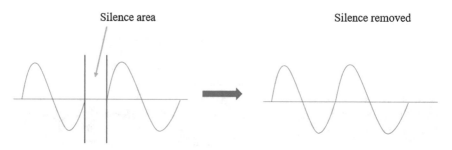

Fig. 12.6 Silence removal

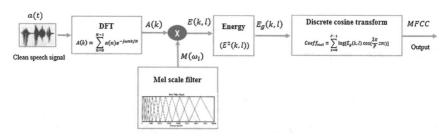

Fig. 12.7 Procedure for MFCC feature extraction

• *Framing*

After silence removal samples obtained having unequal length, to maintain sample size framing is applied. In previous case as speech is nonstationary signal, it's difficult to deal with this signal in continuous domain. To make the signal stationary, split the signal in the interval of 5 s each and apply windowing in each sample to make small samples of milliseconds to extract the features without loss of information due to aperture effect and discontinuities [29].

• *Feature Extraction*

Features are important since the information they produce are the backbone of model for classification purpose. mel frequency cepstral coefficient (MFCC) is the most relevant and effective feature in the field of speech processing [30]. MFCC extracts noncorrelating properties of cepstrum and also compensates channel distortion. Speech sample a(t) is transformed to DFT sample A(k) and windowed by mel scale $M(\omega_l)$. These are further squared to calculate energy. Logarithm of this energy is taken with its discrete cosine transform to get MFCC feature vector. Block diagram for MFCC calculation is shown in Fig. 12.7.

After framing of signal, the discrete Fourier transform (DFT) is calculated. This is important for calculation of energy of the speech signal. Periodogram is an important measure which represents the list of frequencies present in the speech. This function is similar to cochlea of human ear which vibrates and the brain understands the frequency with the help of vibration of different areas of cochlea.

For automatic speaker recognition individual frequency is not important, so a set of frequency named bins is used to calculate the average energy present in the speech. The formation of bins is based on mel scale or mel filterbank. The formula for mel scale is:

$$M(f) = 1125 \, \ln \left(1 + \frac{f}{700} \right)$$

In above equation ln has importance of its own because it is a channel normalization technique which allows cepstral mean subtraction [27].

- *Training and Classification*

The feature extracted from the method discussed above is further used in training. From the abovementioned procedure 13-dimensional feature vector has obtained for each class.

Traditional classification method such as SVM for multiclass classification along with k-NN with its variant is used for classification. Classification using k-NN is a lazy learning approach [31]. k-NN represents each sample point in V-dimensional space where V shows the number of features. In this experiment fine k-NN uses one neighbor with equal weight, while medium k-NN uses 10 neighbor, and coarse kNN takes 100 neighbor, equal weight, and Euclidian distance measurement for classification algorithm. In weighted k-NN instead of equal weight squared inverse weight is used with 10 neighbors and Euclidian distance calculated via MATLAB 2016b.

Principle of Nearest Neighbor Classification Approach
The kNN is a proximity-based lazy learner. This classifier is flexible to find training instances that are comparatively similar to test instances. A k-nearest neighbor classifier characterizes each instance as a data point in V-dimensional space where V is the number of features. By having test examples, proximity is computed to the rest of the instances in the training set by using proximity measures [32]. The algorithm for nearest neighbor classification is presented below.

Ensemble models are combination of multiple models created for improved results. Random subspace method (RSM) based on randomized search improves the accuracy but has drawback that randomly decided features cannot guarantee useful information. k-NN is very sensitive to subspace selection; hence a method reduces the error of non-discriminant information. The method subspace is not chosen randomly but selected as the points which reduce the weighted error rate in each boosting step [31]. Algorithm for ensemble subspace k-NN is shown in text box.

12.7 Results

This paper presents a method of speaker recognition and gender identification for bilingual speech signal. To examine the accuracy of speaker and gender recognition, various tests have been performed for classification with different classifiers. The classifiers used are SVM, coarse kNN, medium kNN, fine kNN, ensemble kNN, and weighted kNN. Table 12.1 shows the result in terms of classification accuracy for different classifiers.

From the above table, weighted k-NN performance is best among all the classifiers. Ensemble method is said to be more accurate but this gives less accuracy than others for current dataset. The area under the curve for gender classification and speaker recognition is given in Fig. 12.8. The AUC for gender recognition is 0.98 for weighted k-NN and it is 0.97 for speaker recognition with same classifier.

AUC is the area of ROC curve drawn between false-positive rate and true positive rate obtained from the classifier. The AUC is the measure of evaluation for the proposed model. If the model is perfect AUC would equal to 1 and if the model performs random guessing AUC would be 0.5. With reference to Fig. 12.8 it can be seen that the proposed model has AUC near to 1 which shows the better performance of the model using weighted kNN classifier.

Comparative analysis of different classifiers is shown in Fig. 12.9. The comparison can be discussed in following points:

- The bar chart represents higher classification accuracy for gender recognition than the speaker recognition problem.
- Recognition performance for Hindi utterances for both the cases showed maximum accuracy than classical recognition for English utterances.
- Mixed utterances gave lesser accuracy for both the cases that emphasize on the effect of language on classification.

Table 12.1 Classification accuracies with different algorithms

Classifiers Data	SVM	Coarse kNN	Medium kNN	Fine kNN	Ensemble kNN	Weighted kNN
Gender classification English utterances	82.7	86.5	91.8	92	91.8	**92.6**
Gender classification Hindi utterances	87.1	89.7	93.5	93.9	93.4	**94.2**
Gender classification mixed utterances	84.3	86.4	91.4	91.6	90.7	**92.2**
Speaker recognition English utterances	78.9	71.3	81.7	82.2	82.9	**83.5**
Speaker recognition Hindi utterances	81.2	76.5	86.3	87.5	88	**88.1**
Speaker recognition mixed utterances	76.7	71.1	81.2	82	81.9	**83.2**

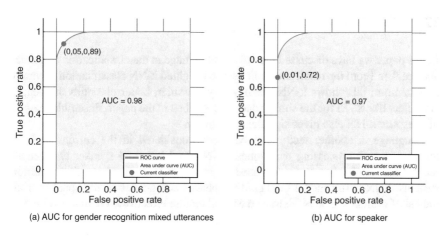

(a) AUC for gender recognition mixed utterances

(b) AUC for speaker

Fig. 12.8 AUC obtained for weighted k-NN. (**a**) AUC for gender recognition mixed utterances. (**b**) AUC for speaker recognition mixed utterances

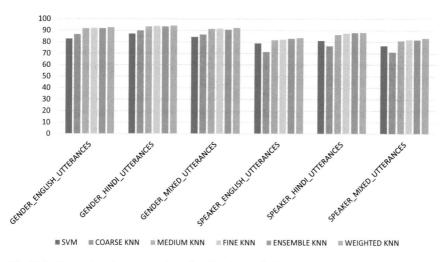

Fig. 12.9 Comparison between various classification methods

- Weighted kNN provides better results for all combination.
- Conventional SVM classifier for speaker recognition gives least accuracy for the present scenario of different languages while fine and coarse kNN have significant performance.

Comparison between various classification methods shown in bar graph (Fig. 12.9).

12.8 Conclusions

In this paper we have discussed the effect of language in the classification of gender and speaker. From the result shown in Sect. 4 weighted k-NN classification accuracy is maximum; this shows k-NN, itself a lazy classifier, better classifies than eager classifiers like SVM for the similar dataset described in the paper. Ensemble method of subspace k-NN also gives significant accuracy.

Language is another factor that has been considered in the current work to improve speech processing techniques. kNN-based bilingual speaker and gender recognition model has been proposed to increase the versatility of recognition systems. kNN-based model predicts the speaker and gender of the speaker from the list of present speakers. Extracted MFCC feature effectively creates envelope of vocal tract. Experimental results showed that weighted kNN outperforms than other classifiers for both speaker and gender recognition. Weighted kNN classification accuracy is maximum; this shows kNN, itself a lazy classifier, better classifies than eager classifiers like SVM for the similar dataset described in the current work.

From Table 12.1, it is seen that language affects the classification performance of both gender and speaker, speaker recognition performance more affected than gender recognition. Since the speakers and background environment are same for both Hindi and English recording sequence, classification performance degrades due to different handsets used for recording.

Future scope of the presented work can be in the direction of improving the performance of speaker and gender recognition. The approach can be extended to identify speakers with the presence of multiple factors such as noise, emotions, and language simultaneously. Multilingual dataset can also be taken into account for the current work's future direction.

References

1. B. Gold, N. Morgan, *Speech and Audio Signal Processing: Processing and Perception of Speech and Music* (Wiley, New York, 1999)
2. J.P. Campbell, Speaker recognition: a tutorial. Proc. IEEE **85**(9), 1437–1462 (1997)
3. W. Yuan, B. Xia, A speech enhancement approach based on noise classification. Appl. Acoust. **96**, 11–19 (2015)
4. N. Singh, R.A. Khan, R. Shree, Applications of speaker recognition. Procedia Eng. **38**, 3122–3126 (2012)
5. P. Rose, Technical forensic speaker recognition: evaluation, types and testing of evidence. Comput. Speech Lang. **20**(2–3 Special issue), 159–191 (2006)
6. A. Khosravani, M.M. Homayounpour, A PLDA approach for language and text independent speaker recognition. Comput. Speech Lang. **45**, 457–474 (2017)
7. H. Beigi, *Fundamentals of Speaker Recognition* (Springer US, Boston, 2011)
8. M.F. Abdollah, M.N. Kamarudin, H.N.M. Shah, M.Z. Ab Rashid, C.K. Lin, Z. Kamis, Biometric voice recognition in security system. Indian Journal of Science and Technology **7**(2), 104–112 (2014)

9. C.S. Kumar, P.M. Rao, Design of an automatic speaker recognition system using MFCC, vector quantization and LBG algorithm. Int. J. Comput. Sci. Eng. **3**(8), 2942 (2011)
10. M.A. Grasso, Automated speech recognition in medical applications. M.D. Comput. Comput. Med. Pract. **12**(1), 16–23 (1995)
11. K. Bain, S.H. Basson, M. Wald, Speech recognition in university classrooms, in *Proceedings of the Fifth International ACM Conference on Assistive Technologies – Assets'02*, 2002, p. 192
12. J. Li, L. Deng, R. Haeb-Umbach, Y. Gong, *Robust Automatic Speech Recognition: A Bridge to Practical Applications* (Academic Press, Amsterdam, 2015)
13. F. Saki, N. Kehtarnavaz, Background noise classification using random forest tree classifier for cochlear implant applications, in *ICASSP, IEEE International Conference on Acoustics, Speech and Signal Processing – Proceedings*, 2014
14. N. Mohammadiha, P. Smaragdis, A. Leijon, Simultaneous noise classification and reduction using a priori learned models, in *IEEE International Workshop on Machine Learning for Signal Processing, MLSP*, 2013
15. A.F. Martin, C.S. Greenberg, NIST 2008 speaker recognition evaluation: performance across telephone and room microphone channels, in *Proceedings of the Annual Conference of the International Speech Communication Association, INTERSPEECH*, 2009, pp. 2579–2582
16. N. Cvijanović, P. Kechichian, K. Janse, A. Kohlrausch, Effects of noise on arousal in a speech communication setting. Speech Commun. **88**, 127–136 (2017)
17. A. Mencattini, E. Martinelli, F. Ringeval, B. Schuller, C. Di Natale, Continuous estimation of emotions in speech by dynamic cooperative speaker models. IEEE Trans. Affect. Comput. **8**(3), 314–327 (2017)
18. S. Gold, Voice biometrics: real-world issues and solutions. Biometr. Technol. Today **2010**(5), 6–7 (2010)
19. O. Nehru, K. Kumar, Review paper of voice biometrics. Int. J. Eng. Sci. Comput. **6**(5), 5257–5260 (2016)
20. S. Chu, S. Narayanan, C.C.J. Kuo, Environmental sound recognition with time frequency audio features. IEEE Trans. Audio Speech Lang. Process. **17**(6), 1142–1158 (2009)
21. F. Richardson, S. Member, D. Reynolds, N. Dehak, Deep neural network approaches to speaker and language recognition. IEEE Signal Process. Lett. **22**(10), 1671–1675 (2015)
22. R. González Hautamäki, M. Sahidullah, V. Hautamäki, T. Kinnunen, Acoustical and perceptual study of voice disguise by age modification in speaker verification. Speech Commun. **95**(March), 1–15 (2017)
23. G. Liu, J.H.L. Hansen, An investigation into back-end advancements for speaker recognition in multi-session and noisy enrollment scenarios. IEEE/ACM Trans. Audio Speech Lang. Process. **22**(12), 1978–1992 (2014)
24. S. Safavi, M. Russell, P. Jančovič. Peter, Automatic speaker, age-group and gender identification from children's speech. Comput. Speech Lang. **50**, 141–156 (2018)
25. Z. Tang, L. Li, D. Wang, R. Vipperla, Collaborative joint training with multitask recurrent model for speech and speaker recognition. IEEE/ACM Trans. Audio Speech Lang. Process. **25**(3), 493–504 (2017)
26. E. Lleida, L.J. Rodriguez-fuentes, Speaker and language recognition and characterization: introduction to the CSL special issue. Comput. Speech Lang. **49**, 107–120 (2018)
27. A. Maurya, D. Kumar, R.K. Agarwal, Speaker recognition for Hindi speech signal using MFCC-GMM approach. Procedia Comput. Sci. **125**, 880–887 (2018)
28. N. Upadhyay, A. Karmakar, Speech enhancement using spectral subtraction-type algorithms: a comparison and simulation study. Procedia Comput. Sci. **54**, 574–584 (2015)
29. P. Nguyen, D. Tran, X. Huang, D. Sharma, Automatic speech-based classification of gender, age and accent, in *Lecture Notes in Computer Science (including subseries Lecture Notes in Artificial Intelligence and Lecture Notes in Bioinformatics)*, vol. 6232 LNAI (2010), pp. 288–299

30. S. Sremath, S. Reza, A. Singh, R. Wang, Speaker identification features extraction methods: a systematic review. Expert Syst. Appl. **90**, 250–271 (2017)
31. N. García-Pedrajas, D. Ortiz-Boyer, Boosting k-nearest neighbor classifier by means of input space projection. Expert Syst. Appl. **36**(7), 10570–10582 (2009)
32. T. Pang-Ning, M. Steinbach, V. Kumar, Introduction to Data Mining (2006)
33. http://practicalcryptography.com/miscellaneous/machine-learning/tutorial-spectral-subraction/

Chapter 13
Effective Security and Access Control Framework for Multilevel Organizations

Ei Ei Moe and Mie Mie Su Thwin

13.1 Introduction

13.1.1 Introduction

The relevant data is a valuable and vital asset for most commercial organizations. Today, the business competitions among business organizations are highly raised so data assets must to be prevented from unauthenticated access and malicious operations. Access control is a principal concept in understanding computer and network security and access privacy to protect data, intellectual property, physical equipment, and systems from accident or intentional damage. One of the technologies that organizations have used to achieve this is access control. By making information resources accessible to only authorized users, the mechanism ensures that only information is always available to those permitted to access it. In large and complex multilevel organizations, it allow users with many clearance, authorization, and need to know ability to ensure that accessing given resources or information in system without risk of compromise. In MAC mechanisms, given resources are categorized with a specific classification level, and each user is specified a certain authorization level.

E. E. Moe (✉)
University of Computer Studies, Yangon, Myanmar
e-mail: eieimoe@ucsy.edu.mm

M. M. S. Thwin
Cyber Security Research Lab, University of Computer Studies, Yangon, Myanmar
e-mail: drmiemiesuthwin@ucsy.edu.mm

© Springer Nature Switzerland AG 2019
G.R. Sinha (ed.), *Advances in Biometrics*,
https://doi.org/10.1007/978-3-030-30436-2_13

13.1.2   Objectives of the Proposed System

Objectives of the proposed framework are as follows:

- To observe the security control policies for information management systems.
- To apply the security mechanisms for internal and external threats in a complex multilevel organizations.
- To use accurate access control mechanisms and policies depending on the nature of organizations.
- To propose an effective security framework for multilevel organizations.
- To implement secure and trusted information system by concentrating on aspects of computer security attacks and vulnerability.

13.2   Related Work

Database security means system, processes, and procedures which prevent the organization's database from unintended and unknown actions [1]. The authorized individuals or processes can make authenticated misuse, malicious attacks, or inadvertent mistakes. Fundamentally, there are two types of database security, DAC and MAC also called multilevel security (MLS). In this paper we will describe how to control banking database system by using MLS techniques. In the case of MAC, each subject is given a certain clearance level, and each object is labeled with a certain classification. The specific object can be accessed only by authorized users with right clearance level. Our system is intended to provide right access control based on user's roles that are assigned according to the enterprise's policy decision. Users who own right access account can manage data from database server.

The system is intended to provide right access control based on user's roles that are assigned according to the enterprise's policy decision. In the system, the administrator can access all data and can make all transactions of the whole system and the data occupation of the respective level. The system users have project manager (level 4), assistant manager (level 3), team leader (level 2), and developer (level 1).These levels are defined by the system administrator. User who owns right access account can manage data from database server [2].

Database security and integrity are vital aspects of an organization's security posture [3]. Database security is the system, processes, and procedures that protect a database from unintentional actions. It is important to develop a security model and policies for every system that use database to store data. The various different security models can solve many security problems. All security models' aim is to outline a system authorized and unauthorized conditions and to constrain the system to move into an unauthorized and unsafe state. Security models of the implemented system can rely on either mandatory or discretionary access control mechanisms. In this paper, the main intention is to implement Biba's Ring Policy that is used to maintain integrity of resources and to provide right access control based on user's roles and attributes that are assigned according to UCSY's attendance policy decision in online attendance marking system – a system that will replace paper-based attendance into digital attendance system.

13.3 Background Theory

Computer security is concerned with five aspects:

Confidentiality: It means that sensitive data or information belonging to an organization or government should not be accessed by or disclosed to unauthorized people. Such data include employees' details, classified military information, business financial records, etc.

Integrity: This means that data should not be modified without owner's authority. Data integrity is violated when a person accidentally, or with malicious intent, erases or modifies important files such as payroll or a customer's bank account file.

Availability: The information must be available on demand. This means that any information system and communication link used to access it must be efficient and functional. An information system may be unavailable due to power outages, hardware failures, and unplanned upgrades or repairs.

Authenticity: It is the confirmation of the identity of the user. It is accomplished through something only the user knows, i.e., password; using what the user has, i.e., badge and smartcard; and something that the user is, e.g., biometric analysis, i.e., finger prints, voice recognition, retina, face recognition, etc.

Non-repudiation: It is the assurance that someone cannot deny the validity of something. It offers the ability to decide whether a specified individual took a certain accomplishment such as message sending, information generating, or approving and message receiving.

13.3.1 What Are External and Internal Threats?

Everything or everyone that can cause risks to computer system of organization, computing resources, users, or data owned by business is called threats.

External threats are initiated from an outsider of an organization, mainly from the environments in which the organization makes their operations. The attackers perform various attacks by stealing credentials of a legitimate. Examples of external threats are hacking, code injection attacks, malware, phishing, corporate espionage/competitors, and business partners/contractors.

The second threats initiate from inside the organization. The employees or providers to whom work is outsourced are the key contributors for causing internal threats. Frauds, exploitation of information, damage of information, and sensitive data leakage within organization are the main threats for that organization. Mostly, the employee in every organization can be biggest threats rather than hackers outside the organization.

13.3.2   Security Control Model

An organization should define its security strategies and plans. An organization can use a security model to support the workplace policies or IT security guidelines to be applied in an organization's computer system.

Security Policy
A security policy is a specialized paper which outlines how to protect the organization from threats, including types of computer security threats, and how to handle situations when they do occur. It manages a set of security procedures and purposes desired by an organization. The security policy must identify all of a company's assets and resources as well as all the potential threats to those assets.

Security Model
A security model is a framework in which the security policy is developed for the security needs of organizations. The security control models are used to outline how security patterns will be implemented, what users can access the system, and which information they will have read or write. Essentially, they are a way of defining security policies between users and organization. Security models are generally implemented with integrity, confidentiality, and other essential security controls.

13.3.3   Bell-LaPadula Model

The security model was created for preventing access to objects (data) in a system. It is the first mathematical model that applied a multilevel security policy that is used to express a secure state machine concept.

The properties of this model are as following:

- Simple security property mentions that a given subject at a security level cannot read objects that exist in at a higher security level.
- Star property mentions that a given subject in a security level cannot write objects to a lower security levels.
- Strong star $(_*)$ property mentions that a subject cannot read or write to objects at higher and lower levels.

A problem of this model is it does not perform the integrity of data, although all mandatory access control systems are based on the Bell-LaPadula model.

13.3.4 Integrity

The integrity denotes that the trust worthiness of data or business assets. It plays an important role in IT security and defined the processes of preventing improper and unauthorized changes to the data. Although governmental entities are usually concerned with confidentiality, other business and education organizations might be more focused on the integrity of information.

Generally, four main goals of the integrity are:

1. Prevention of making modifications to data or programs by unauthorized parties.
2. Prevention of making improper or unauthorized modifications by authorized parties.
3. Maintaining external and internal consistency of programs and data.
4. Reflecting the real world by ensuring transition to use data accurately.

13.3.5 Levels of Integrity

Levels of integrity are labels which consist of two parts:

1. Level of Classification
   The form of classification is a hierarchical set.
   Crucial, important, and insignificant can be divided as the example of classification. The highest level is crucial and insignificant as the lowest level. For this case: crucial > important > insignificant.
2. Set of Categories
   This is a compartment which contains label that can be a subset of system's all the sets. The nonhierarchical form is the form of a set of categories.

All subjects and objects in the system give integrity levels. The integrity levels become a dominance relationship between subjects and objects in system. The integrity label corrects the confidence level that may be retained in the data.

13.3.6 Strict Integrity Policy

This policy is an integrity policy and involves among the types of mandatory access control. It is also the dual of Bell-LaPadula model and basically has the following three defining properties.

The *simple integrity property* means the subject can only read objects at its integrity level or above level. The subjects can read objects only if i (o) is greater or equal to i (s).

The *integrity* property* means the subject can only write objects at its integrity level or below level. The subjects can write to objects only if i (o) is less than i (s).

A subject with low integrity level is prohibited from invoking or calling up a subject with a higher level of integrity is called *invocation* property.

13.3.7 Strict Integrity Access Control Model

In the meantime it is an access control policy, it can be represented as the access control matrix. Assume that H (high level) > L (low level) > VL (very low level) are hierarchical integrity levels (Fig. 13.1).

13.3.8 Conditions of Integrity

- Simple integrity condition is also known as "no read down" axiom (Fig. 13.2).
- Integrity star property is also known as "no write up" axiom (Fig. 13.3).

13.3.9 Levels of Entity to Control Security

A reliable computer system ensures that preventing access to objects based on the sensitivity (label) of the information contained in that objects and the proper authorization (clearance) of subjects to get given access to specific objects in the system. A subject is an active entity that makes request to objects and an object is a passive entity that consists of information. Security labels denote the security sensitivity level of:

Fig. 13.1 Access control matrix between subject and object

Subjects	Level	Objects	Level
S1	(H2{a,b,c})	O1	(H2{a,b,c})
S2	(L,{})	O2	(L,{})
S3	(L,{a,b})	O3	(L,{a,b})
S4	(VL,{a,b})	O4	(VL,{a,b})

	Object 1	Object 2	Object 3	Object 4
S1	W	W	W	W
S2	R	R,W	R	W
S3	R	R,W	-	W
S4	R	R	R	W

R - Read Access, W - Write Access

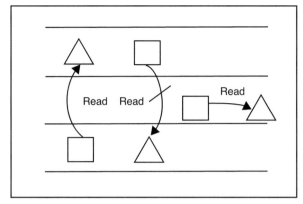

Square - subject , Triangle - Object

Fig. 13.2 Simple Integrity Condition

Fig. 13.3 Integrity star property

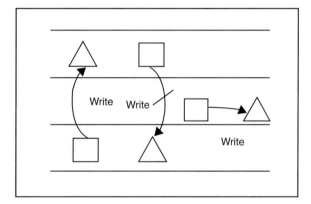

Square - Subject, Triangle - onject

- Subjects that are given clearances labels.
- Objects that are given classifications labels.

Clearance label is considered as security level that an individual user can access the data. This is usually related to a "need to know" necessity. When the clearance and classification labels work together, a user's clearance is a restriction to the access of resources based on their classification.

Each desired security control level is supposed to govern itself and all others below it in this hierarchy structure.

13.4   Access Control

It governs who can access, place, or use what resources in a computing atmosphere and is also an essential component in security. It minimizes the risks of unauthorized access for businesses or organizations. Access control ensures that an acting authenticated user can access only what they are authorized to and nothing more. Generally, access control involves user identification, authorization, authentication, accountability, and audition concepts.

13.4.1   Identification

Identification defines a method to ensure that a subject (e.g., user, program, or process) is the entity it claims to be. It can provide using valid identity such as username or account number and password.

13.4.2   Account Authentication

The authentication is the most elementary requirement for handling of user access in the system. The user authentication comprises the user is who he claims to be actually. In the user authentication process, the subject is always required to enter a second part to the credential set such as password, personal identification number (PIN), or cryptographic key.

13.4.3   Authorization

The subject is actually authenticated when the credentials of identification and authentication match the stored information in database of the system. The system tries to control the access to resources for subject after the subject is authorized.

13.4.4   Accountability

Accountability mechanism keeps track of subject actions in the system. It keeps track of who, when, and how the subjects access the system. It helps in identifying authorized and unauthorized activities between the subject and object.

13.4.5 Types of Access Control

The organizations can apply diverse access control models that depend on their business policies or compliance need and the security levels of organization they want to guard. The core access control types are:

Mandatory access control (MAC) sets access rights by a system administrator (central authority) based on multiple levels of security. This mechanism assigns all resource objects with security labels in the system. These security labels are divided into two information parts: a classification (secret, confidential, unclassified, etc.) and a set of category (the management level, department, or project).

In *discretionary access control (DAC)*, administrators of the protected system set data or resources with the specified policies in which who or what gives authorization to access them. It allows individual user to control access to their own operations and data.

Role-based access control (RBAC) is also known as non-discretionary mechanisms. This mechanism limits access to resources based on individuals or groups with specific functions rather than the identities of individual users. It gives permissions to particular roles that have assigned by the users in an organization.

Rule-based access control model in which the system administrator defines a set of security rules allows or denies access to specific objects by subjects. These often depend on circumstances such as time of day or location.

Attribute-based access control (ABAC) is known as a policy-based access control that evaluates a set of policies, rules, and relationships by using the attributes of users, systems, and environmental situations to manage access rights for users. A key difference among these mechanisms is the concept of policies precise a complex Boolean rule set that can evaluate many different attributes.

13.5 Biba Security Model

Although many governments are principally concerned with confidentiality, most businesses wish to ensure that the integrity of the information is protected at the highest level. When the protection of integrity is vital, Biba is the model of choice by most organization. Bell-LaPadula model cannot grantee data integrity but can offer confidentiality of data. So, this integrity model addresses the requirement of enforcing integrity for such computational environment.

13.5.1   Access Modes of Biba Model

The Biba Model involves the type of access modes. Although these modes of access use different definitions to express them, they are similar to those used in other models. The access modes that can support Biba Model are:

1. Modifying mode permits writing to an object by a subject and is similar to the write mode of another models.
2. Observing mode permits reading to an object by a subject and is a synonym with the read command of other models.
3. Invoking mode permits a subject to interconnect with another subject.
4. Executing mode permits executing an object by a subject. The command essentially allows executing a program which is the object by a subject [4].

13.5.2   Policies of Biba Model

The Biba Model can be separated into mandatory and discretionary policies as two types of policies according to the need of the system.

Mandatory policies in Biba Model are:

1. Strict Integrity Policy.
2. Low-Water-Mark Policy for Subjects.
3. Low-Water-Mark Policy for Objects.
4. Low-Water-Mark Integrity Audit Policy.
5. Ring Policy.

Discretionary policies in Biba Model are:

1. Access Control Lists.
2. Object Hierarchy.
3. Ring Policy.

Biba Model also uses labels to define security. The labeling technique of Biba Model is used to provide integrity levels for the subjects and objects in the system. The labeled objects with a high level of integrity will be more sensitive and accurate than the labeled objects with a low level. The integrity levels are used to restrict the unsuitable modification of data and allow the right operations on that data.

13.5.3   Advantages and Disadvantages of Biba Model

There are many pros in using Biba Model in organizations' security. The Biba Model is simple and can be implemented easily. This model also can provide many different policies based on the different organizations' nature and necessities.

The Biba Model also has some disadvantages. The first issue is that the programmers need to use the right policy and rules according to the implementation of different organizations' security. The second disadvantage is that the Biba Model does not perform about confidentiality, while the Bell-LaPadula can enforce.

Moreover, the Biba Model does not support the granting and revocation of authorization. So access control mechanisms can achieve this failing mechanism. The last disadvantage is that to use this model, all computers in multilevel organizations must provide labeling of integrity for both subjects and objects. There are problems in using Biba Model in the network environment because the labeling techniques cannot be supported by network protocol.

13.6 Security Controls

The security safeguards and security controls in such information security system are capable of several criteria such as preventing security incidents, minimizing risks, detecting attackers, and recovering damage to normal state. For multilevel organizations, it must consider the effective security control mechanisms to become more secure and safe. A number of different types of user such as unknown guest users, administrative users, and regular authenticated users may be consisted in multilevel organization's environment. The different set of data and resources are permitted to access by many users. It will discuss the following necessary security controls for information security management system of that organization.

13.6.1 Account Authentication Control

This is the handling of user request to the system and is also the most basic dependency. The user authentication involves proving that a user is in fact who he or she claims to be. If there is no such facility, the system will assume all users as unknown which is the lowest possible level of trust. Most systems use the simple authentication method in that the user submits username and password for checking account validity [5]. But this proposed system uses two-factor authentication mechanism to avoid SQL infection attack.

13.6.2 Handling User Access Control

It is making to impose correct decisions about whether each individual user request should be allowed or denied in the process of handling user access. If this is functioning correctly, the system detects the identity of the user from whom each request is received [5]. Access control mechanisms can support the need to know restriction. The need to know ability ensures that only authorized users gain access to information or systems necessary to accept their responsibilities. To approve these control mechanisms, access control is used as basic theory.

13.6.3   Using User Input Control

The users in most organizations often have not been potentially aware of the risky faults they're making. The documents are put onto unsecure cloud apps, to working from home on their personal devices, untrusted user input into application and even sharing passwords. It can then somehow be made vulnerable to information theft and data leakage within organizations. So, user input handling controls basically prevent the effects of mistakes made by nontechnical users through a secure work culture. As a result of this control in desired proposed system, input sanitization method can be used.

13.6.4   Handling Communication and Data Transfer Controls

The employees within an organization can communicate safely to each other according to access control security features. These features control how users and system communicate and interact with other systems and resources.

13.6.5   How to Handle Employees' Daily Operations Controls?

The employees in organization have specific individual role and responsibilities to perform business operations and duties. Before using the proposed security framework, unique passwords for different logins are identified by system administrator. To check improper activity, employee's usage record will traced by auditing method. When transferring sensitive data of organizations, the need to know the level of employees will determine the access of these types of data. While an employee requests the data, process, or device of organization, respective security controls examine whether to accept or not deny each request.

13.7   What Is Multilevel Organization?

The classification of organizations is defined according to a hierarchy of authority and different responsibilities of individual employee. The three management levels most organizations have are first-level, middle-level, and top-level employees. The management style is influenced by the goals and purpose of the organization.

The term multilevel stands from the security classification of the defense community with confidential, secret, and top secret clearance levels. Individual users must be approved with accurate clearance levels before they can access the set of classified information. The confidential clearance users are only authorized to view confidential documents; they are not reliable to look at secret or top secret information. The multilevel organizations include many user classification levels and set of categories of business's digital resources [6]. A multilevel system is a single computer system that handles various classification levels between subjects and objects. In this system, access rights are associated with user, and roles are granted to appropriate user.

13.8   Secure Framework for Multilevel Organizations

13.8.1   Secure Framework for Multilevel Organizations (SFMO)

Secure framework for multilevel organizations (SFMO) is a security framework that involves the integration of security mechanisms such as user identification, account authentication, user authorization, access permission, user classification and access privacy, data classification in organization, and protection of most sensitive data. The goal of SFMO is to defend the vulnerability of insider and external threats, to protect from stealing company's business legal resources, and to secure any different structure of user levels and system resources in an organization. The following figure is the pattern of SFMO which makes a secure and safe environment for multilevel organization (Fig. 13.4).

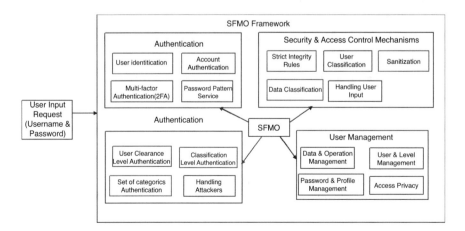

Fig. 13.4 Components of secure framework for multilevel organization

13.8.2 Features of Secure Framework for Multilevel Organizations (FSFMO)

Organizations with many staff management levels are generally called multilevel organizations which include groups of employees with different authority levels, many operations or functions, and relevant companies' data. These complex natured organizations must have an effective security countermeasure to keep their business safe from cyber threats. So, a secure framework will develop with the deployment of a set of security controls which obey the computer security aspects. The proposed secure framework for multilevel organizations will support the following features:

User Identification

It ensures a valid identity for an individual in the organizations with username and user level, for example, the system administrator login with username "admin" and user-level "administrator."

The rules of the authorized user and anonymous user are as follows:

```
IF username_identity is exist AND userlevel_ identity is exist
      THEN GRANT
         "User Identification" request is valid.
ELSE IF username_identity does not exist AND userlevel_identity
is unknow level
         THEN GRANT
         "User Identification" request is invalid and return invalid
         message.
END IF
END IF
```

Account Authentication with Two-Factor Authentication

This component of feature checks the identity of an individual user with password or PIN after user identification process is valid. The rules of the account authentication are as follows:

```
IF username is valid AND userlevel is exist AND password is
correct
      THEN
         "Authentication" is successful and the user is authenticated.
 END IF
ELSE
      THEN
         "Authentication" fail.
END IF
END IF
```

The single-step logins will compromise an attacker to get the usernames of other users, password prediction, and exploiting defects by bypassing the login functions. Therefore, the proposed framework uses two-factor authentication (2FA) to secure the system. The 2FA, two-step verification, is an extra security control that is also known as "multifactor authentication" (MFA). This control can be used as a combination of something the users have and something the users know.

User Authorization

This feature controls the access to system resources for authorized and unauthorized users. The user is authenticated and gets fully access right according to the user level in system. For example, the guest can view the profile of the university, while the authorized student user can view and search the respective information of that university.

The rules of the authorization are as follows:

```
IF userlevel is Level 1 AND classification level is Top secret
     THEN GRANT
     Give "Level 1 User" access right and can request to access
     "Top Secret object".
END IF
```

Integrity and Access Control

This type of feature applied Biba Model's Strict Integrity Policy. The policy provides subjects and objects with the integrity levels. $L = (C, S)$ will represent as integrity level. In this equation, integrity level is defined as capital letter (L), the classification label is defined as capital letter (C), and set of categories label is defined as the capita letter (S). No write up and no read down are the main function of Strict Integrity Policy.

```
Algorithm 8.1
Integrity and Access Control Algorithm
INPUT: subject Lₙ= (Cₙ, Sₙ), object Lₙ₊₁= (Cₙ₊₁, Sₙ₊₁),
OUTPUT: if (Sₙ ₊ ₁ ⊄ Sₙ) REJECT, otherwise
METHOD:
        1. Initialize Cₙ,Cₙ₊₁
        2. Cₙ and Cₙ₊₁ are assigned with values.
        3. If Sₙ ⊇ S ₙ ₊ ₁:
                    If Cₙ = =Cₙ₊₁: Lₙ WRITE Access to L ₙ₊₁
                        End if
                    Else If Cₙ > Cₙ₊₁: Lₙ WRITE Access to Lₙ₊₁
                        End if
                    Else If Cₙ < Cₙ₊₁: REJECT WRITE Access to Lₙ₊₁
                        End if
                    Else If Cₙ > Cₙ₊₁: Lₙ REJECT READ Access to Lₙ₊₁
                        End if
            End if
        4. return access request type
```

The subject L_1's integrity level $= (C_1, S_1)$, and the object L_2's integrity level $= (C_2, S_2)$. If the classification level C_1 (top secret), C_2 (secret), and S_2 is a subset of S_1, the subject level dominates the object level. According to the Biba's Strict Integrity. Policy, subject writes the object because L_1's integrity level is higher than integrity level of L2.

User Classification and Access Privacy

This feature can define the access rights and responsibilities for each user level and individual user. Access control rule defines the clearance (need to know) level for the user in the system. It provides the user profile management, privacy control for personal information confidentiality, password management, self-service, etc.

The rules of the user privacy and access right are as follows:

```
IF userlevel is Level 1 AND classification level is Top secret
AND set of categories "Group 1"
        THEN GRANT
        Give "Level 1 User" access right.
        Can request to write "Secret object".
        Can connect to everyone in "Group 1"
END IF
```

Data Classification

It classifies all data and resources of the organizations with integrity access control policy. It can provide labeling with classification level and set of categories for system data and resources.

The rules of the data classification are as follows:

```
IF object level is obj 1 AND classification level is Top secret
        THEN GRANT
        Can access to writ with same level subject (user)
END IF
```

Sensitive Data Protection

The feature contains the logic to protect data leakage of very sensitive data and digital asset from company' internal to the outside of organization by data encryption method.

13.8.3   Design of Proposed System

The proposed system aims to ensure that a secure framework of multilevel organizations. There are many levels of user (subjects) and different categories of objects (data) in the system. As the aspect of authenticity, this system identified the users by the predefined username and password before they enter the system. Users must enter username and password to check the validity of member in the system. So, the data availability is depending on the authentication checking. The system authorized users if their login name, password, and user level is correct. The user performs the operations of system resources according to their clearance level that had given by system administrator. Moreover, the system administrator classifies the organization's business data and resources with types of category and important level of business.

At the processing of the objects:

Integrity aspect:

- Higher-level subject can write to the lower level object.
- Same level subject can write to the same level object.
- Subject at lower level can't write to the object at higher level.
- Higher-level subject can't read to the lower level object.

As the system permits the access of each right user, the whole system can also retain the confidentiality on each object. And it can also comprise the protection of sensitive information of the organization in this framework (Fig. 13.5).

13.8.4   System Flow of Proposed System

The flow of this system is designed with only four user classification level as an example (Fig. 13.6).

13.8.5   Case Study for Multilevel Organizations

The following are three case studies for multilevel organizations and their user-level classifications.

1. *MCC Training Institute*

The first case study of multilevel organization is an education training center. There are three departments in that organization such as education services, finance department, and human resource department. In education services, it is divided into two teams: management and teaching.

The management includes CEO, COO, GM, dean manager, and AGM. At the teaching team, there are professors, lecturer, assistant lecturer, and class tutor, respectively.

So the following can show the user classification level of teaching team in MCC Training Institute. The highest user level is professor, and the lowest user level is class tutor. Each user level can have individual user right and responsibility. And the data and resources (object categories) had assigned to them by system administrator (Fig. 13.7).

2. *Mahar Myaing Trading Co. Ltd*

The second case study is a trading company organization. This organization generally includes CEO, manager, assistant manager, accountant, and office staff. CEO is the highest user level in that organization. So he or she can fully access almost system resources and files. The lowest user level, office staff, can request to access with the same or lower level of objects in the organization (Fig. 13.8).

3. *Galaxy Software Company*

The third one is a case study of a software company. In software company, it may have different departments such as development, testing, maintenance, customer service, and sales and marketing if this company is a big software company. However, there are few numbers of the company with many departments. Especially, only one team performs different duties at the same time in small office. So the

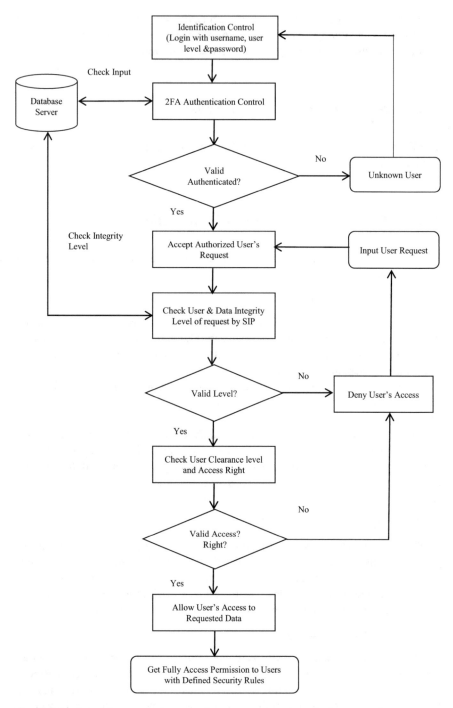

Fig. 13.5 Design of the proposed system when user requests access to data

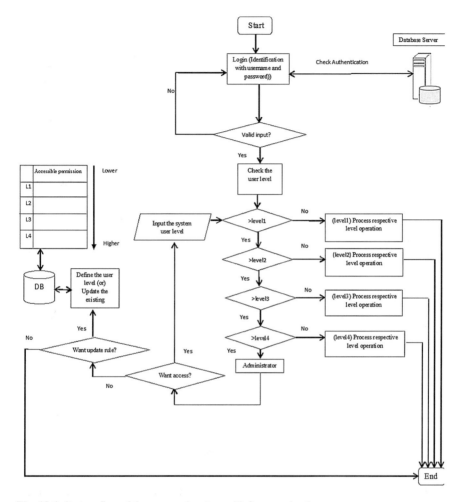

Fig. 13.6 System flow of the proposed system with four user level

company is generally divided into five user level with project manager, programmer, developer, system analyst, software designer, and tester (Fig. 13.9).

The labels of classification and set of categories for subject (user) and object (resource, information, and file) are specified by system administrator (SA). One or more system administrator will control the system. The needs and nature of the organization change user-level classification and data categories.

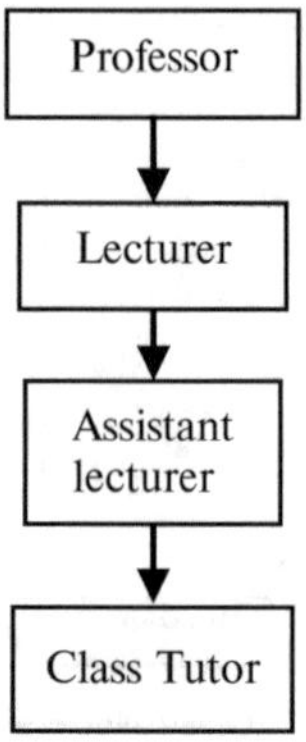

Fig. 13.7 User classification
level of MCC case study

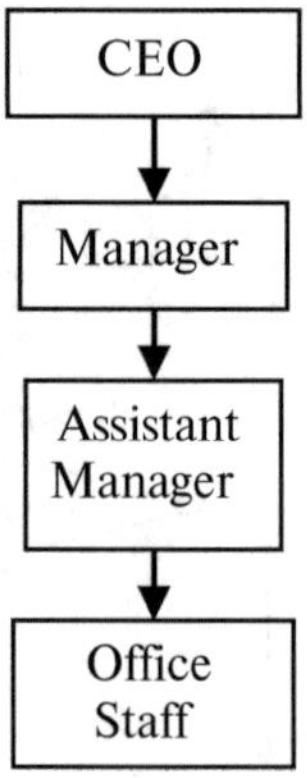

```
┌──────────────┐
│  Professor   │
└──────────────┘
        │
        ▼
┌──────────────┐
│   Lecturer   │
└──────────────┘
        │
        ▼
┌──────────────┐
│  Assistant   │
│   lecturer   │
└──────────────┘
        │
        ▼
┌──────────────┐
│ Class Tutor  │
└──────────────┘
```

Fig. 13.8 User classification
level of trading company case
study

```
┌──────────────┐
│     CEO      │
└──────────────┘
        │
        ▼
┌──────────────┐
│   Manager    │
└──────────────┘
        │
        ▼
┌──────────────┐
│  Assistant   │
│   Manager    │
└──────────────┘
        │
        ▼
┌──────────────┐
│    Office    │
│    Staff     │
└──────────────┘
```

13.9 Privacy Policies at Workplace

The proposed secure framework imposes the multilevel organizations to be more
secure and trusted. The data in the organizations categorize into the sets by defining
labels, and the users define classification level. The system administrator can
manage internal operations for individual users and data. The highest-level user in
the system knows everything and can access any resources. But every user needs to
protect user privacy that they wish to keep and their personality information that may
be confidential. So the organization must identify basic rules for privacy policies for
users.

The following are basic rules for privacy policies to define in organization by
system administrator:

1. It intimates users about what you collect and why and what you will do with it.
2. It should limit the collection of information and collect it by fair and lawful
   means.
3. Inform users about the potential collection, use, and disclosure of personal
   information.

Fig. 13.9 User classification
level of software company
case study

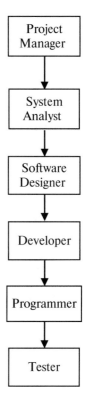

4. Keep user's personal information accurate, complete, and up-to-date.
5. Provide users access to their personal information.
6. It must keep user's personal information secure [7].

13.10   Conclusion

The intellectual property and assets of an organization or a company are mostly
stored as digital format in database. Database security is an essential part in the
information security management of nowadays digital system. The employees in
every organizations can become the biggest threat because they often are not
mindful on the possible risky mistakes whether they are doing or not in the database
of that organization. In this case, the most effective cybercrime are facilitated by
employees' accidental or careless actions. So, the organizations need to address
the problems and solutions of this risk. The protection of database is to restrict
unauthorized employee to access a company's confidential data and digital property.

The proposed secure framework implements the collaboration of access control,
integrity, and possible security control mechanisms to ensure that defend malicious

threads, end user's faults and intentional attacks of information stealers. This proposed framework ensures that the organizations get a security-driven work culture and acts like a defensive wall of the organization's information security system.

References

1. Z. Aung, Database server security for banking information system. M.C.SC Dissertation. University of Computer Studies, Yangon, 2010
2. M.K. Moe, Implementation of mandatory access control using staff levels. M.C.Sc Dissertation. University of Computer Studies, Yangon, 2016
3. S. Zune, Mandatory access control by Biba model. M.C.Sc Dissertation. University of Computer Studies, Yangon, 2018
4. N. Balon, I. Thabet, Biba Security Model. CIS 576. (2004, March 17)
5. D. Stuttard, M. Pinto, *The Web Application Hacker's Handbook: Finding and Exploiting Security Flaws*, 2nd edn
6. Red Hat®, Multi-level Security (MIS), https://www.centos.org/docs/5/html/Deployment_Guide-en-US.html. Accessed 1 May 2019
7. Council, EC, *Certified Ethical Hacker Note*

Chapter 14
Dimensionality Reduction and Feature Matching in Functional MRI Imaging Data

B. Sowmya, Shruti Bhargava Choubey, and Abhishek Choubey

14.1 Introduction

Multivariate pattern analysis, plan classifiers in order to perceive cerebrum institution structures which identified external lifts or social responses, and gives outcomes as a logical standard for getting information encoded in spatial related voxels within f-magnetic resonance imaging [1–5]. Most f-magnetic resonance imaging applications are used in desired issue which attempt to relate inside social event irregularity with direct responses. In any case, whole cerebrum f-magnetic resonance imaging information is of lofty spatial volumes, yet the amount of tests (i.e., subject and exploratory primers) are conventionally obliged, and it particularly hampers the introduction of precedent affirmation techniques. It can be associated with f-MRI to exhibit the association between the view of subjects or test primers and mind sanctioning plans. Estimation decline is essential to extract features from high-dimensional f-magnetic resonance imaging information for higher figure exactness and better model interpretability. Customary measurement decrease frameworks used for f-magnetic resonance imaging information, and the lofty-dimensional information can be applied map into an immediate subspace crossed through some lethargic sections, for instance, foremost segment investigation (PCA) and incomplete least squares (PLS). In any case, these methodologies expect that the accumulated information and their class names are straightforwardly related, which may require physiological assistance and isn't according to the puzzling and intrinsic nonlinear neuro-components of the cerebrum. Thusly, straight systems can't absolutely evaluate the best estimation decline (EDR) headings. Nonlinear coordinated estimation decline techniques, like, bolster vector machine by recursive

B. Sowmya · S. B. Choubey (✉) · A. Choubey
Sreenidhi Institute of Science and Technology (SNIST), Hyderabad, India
e-mail: shrutibhargava@sreenidhi.edu.in; abhishek@sreenidhi.edu.in

© Springer Nature Switzerland AG 2019
G.R. Sinha (ed.), *Advances in Biometrics*,
https://doi.org/10.1007/978-3-030-30436-2_14

component transfer and Gaussian methodology dormant variable mold, can't be planned for the examination of inherited information or else pictures, are worn in neuro imaging revise due to which lofty computational eccentrics failed to be managed.

Now, we hope toward extending a novel controlled estimation decline method, called as rule portion – cut turn around backslide (PCA-SIR), which includes a course of an unproven PCA along with a coordinated SIR, to depict the handy association among pointers (f-magnetic resonance imaging voxels) and their imprints. SIR, which was used to assess EDR orientation, which reduces information clear names under a straight otherwise nonlinear association. Different productive utilization of SIR contain be represented, in bioinformatics. In case, SIR can't oversee conditions like (1) markers are determinedly related; (2) the amount of pointers are significantly greater than the amount of tests, therefore it isn't sensible for getting ready f-magnetic resonance imaging information. To deal with issue, principal componenet analysis used before SIR to orthogonalize and reduce the amount of pointers with the objective that the PCA-diminished markers can satisfy the requirements of SIR. We at first depict the nuances of PC-SIR. Until which, we continue to differentiate PC-SIR and standard methodologies in f-magnetic resonance imaging contemplating reinforce vector backslide and fragmented least square backslide by techniques for diversions containing distinctive utilitarian associations. The new technique is used to perceive the torment linked cerebrum structures, similarly envisioning the element of enthusiastic torment wisdom using an anguish prompting f-magnetic resonance imaging information set [11, 21]. The perceived cerebral structural zones and desire shows are pondered and analysed by the new technique.

Dynamic magnetic resonance imaging (d-magnetic resonance imaging) expects a basic occupation in different clinical applications for thinking about emotional and quantitative components of various physiological activity and body organs functionality. Regardless, d-magnetic resonance imaging's spatio-transient objectives are normally exchanged off in light of unavoidable long information acquiring times. Forefront methods like PLS and PCA in pressed distinguishing (CS) similarly as low-position (LR) depicting solid results in changing dMR pictures from sub-Nyquist tried k-space information, abusing pitiful or conceivably low-position priors along spatio-transient direction. While an enormous segment of sparsity-propelling methods in magnetic resonance imaging revolves around Fourier and wavelet bases and low-position gauge is essentially established on principal composition analysis (PCA) to get the capriciousness of the information, new enhancements in dimensionality decline strategies, called complex learning, which gives better ways to deal with look-in information connections. Unable to withstand the way that a large portion of complex learning techniques are used for information plan and observation, complex based priors have in all regards starting late pulled in premium magnetic resonance imaging multiplication.

Benefitting by complex learning conflicts, this document displays a complex base structure, we name as M-magnetic resonance imaging also inside this structure, and makes two original strategies M-magnetic resonance imagingr1 and M-magnetic resonance imagingr2 used for picture entertainment within d-magnetic resonance imaging as of exceedingly underexamined k-space information. Picture xi, taken from a d-magnetic resonance imaging game plan and embedded in a high estimation space CN, is exhibited as a point on or close to an intricate M with the ultimate objective that M $\subset$ CN. Relations among pictures in d-magnetic resonance imaging are picked up as of suitably portrayed neighbourhoods in M. The neighbourhoods can be predefined Gaussian part that has been exhibited to be viable in showing scattering and warmth guide shapes, we propose here rather an information-driven learning approach where neighbourhoods are portrayed by the information themselves reliant on information expressed by pitiful relative relations. To propel low-dimensional delineations of information, a low-dimensional method can be used to facilitate defends the intricate geometry depicted through the academic neighbourhood be furthermore figured [18].

Dementia is a cerebrum issue that is depicted by a wearisome lessening in mental limit in view of defeat or else trouble to neurons within the brain. The phases of dementia has its own cerebrum issues that result in a ton of symptoms that aggravate commonplace personality limits, like empathizing memory, basic reasoning, and utilization of language, adequately can impact the patients' consistant work. Starting late, the inescapability of dementia has risen awesomely all throughout the world. In England, more than 500,000 people diagnosed with dementia in 2011. This number has by and large extended. Every year 46.8 million natives global detected dementia. The estimate is about a colossal addition of about 75.6 million by 2030, which will essentially altogether increment by 2050 135.5 million. Supreme restorative administrations' considers people with dementia signified further than 1% of the overall absolute national yield (GDP) or US$604 billion, in 2010. Detection of dementia expect a noteworthy activity in perceiving the right treatment, deflecting or thwarting scholarly limit breaking down, and in searching for the reasonable assistance underground bug preparing for what's to come. Regardless, the early finding of dementia is a troublesome errand owing on the way to the unfathomably flightiness of dementia depictions in neuroimaging information [6].

Neuroimages consistently suspected to the negative impacts of the scourge of dimensionality to facilitate and construct them incredibly hard task imagined along with organized within examination arrange. High dimensionality of pictures can be influenced by the introduction of machine learning (ML) classifiers in the midst of setting it up on top of this picture information consequently to facilitate over fitting augmentations and previously the unpredictability of information increases. To tackle the problems of various dimensionality decline frameworks comprise made just before decrease along with keep up a key separation from the special effects of high-dimensionality problem, improve a precision of hypothesis in addition to after that compose AI figuring's logically solid. The work efficiency ponders over the design to diagnose dementia has been realized subject to features removed designed for isolating among multi-periods of the earnestness of Alzheimer's disorder (AD). Usually the component used for isolating among different sorts of dementia is so far a space. The region of interest of our philosophy in the direction of separate type skin texture starting different region of interest used for isolating between conventional controls, solitary patients by auto-encoders and patients with Front fleeting projection dementia (FTD) by using a small auto-encoder (SAE) [26–30]. The planned system is affirmed by using straight discriminate examination (LDA), determined backslide, and convolutional neural network joined with determined backslide (CNN+ backslide) classifiers with both the SAE-based low-dimensional information set and PCA-driven depiction for examination. The planned CAD approach convey the strange state depiction of neuroimages information to isolate the key features that recognize various sorts of dementia. The features be appeared differently within relation to that expelled from the notable Principal Component Analysis within getting ready multi-type dementia course of action along with the exploratory result be analysed. Finally, region V wraps up this examination.

14.2 Existing Works

Interpreting the impression of torment from f-magnetic resonance imaging utilizing multivariate example examination, agony is known to contain material, abstract, and enthusiastic points. Despite different reasons f-magnetic resonance imaging contemplates, regardless, the vestige gives a spatial flow of movement be satisfactory towards encode whether a redesign be viewed like anguishing or else not. During this examination, we separated f-magnetic resonance imaging information based on percept fundamental initiative assignment inside which individuals exhibit on the way to close edge laser beats. By the multivariate examinations lying on dissimilar spatial scales, we investigate the insight further reaches of f-magnetic resonance imaging information designed for translating whether an update has been viewed when anguishing. Our examination yield to conclusion of mind territories: in the midst of anguish desire, development inside the periaqueductal diminish (PAG) along with orbito frontal cortex (OFC) dealt with the mainly exact primer by-fundamental isolation among painful and non-troublesome experiences; while in the midst of the real impelling, basic and discretionary somato material cortex, front insula, dorsolateral along with ventrolateral prefrontal cortex, with OFC be commonly discriminative. The main goal for torment acumen from the affectation time period, regardless, was engaged by the joined development in misery regions routinely implied as the 'torment grid'. Our result facilitate the neural depiction of (close edge) torment be spatially appropriated along with capability of the finest portrayed on a widely appealing spatial scale. Despite its utility in structure-up structure-work mappings, our system bears fundamental by-primer gauges and thusly addresses a phase in the direction of the goal of setting up an objective neuronal marker of torment acknowledgment [9].

The present examination use a multivariate translating advance (i) to anticipate experience of close edge torment starting mind development in the midst of the desire with receiving of torment, and (ii) to assess the flowed thought of torment reading. Our examination provoked three standard disclosures. In any case, we insisted that it is possible to unravel the impression of torment from primer sagacious whole cerebrum f-magnetic resonance imaging information, even without physical redesign changes. Accordingly, we have displayed the nearness of an honest yet basic accurate association between extents of cerebrum development and torment insight. Second, while most cerebrum locale by and large associated with anguish is mulled over above chance decoding, these territories differentiated as towards their farsighted exactness. Our examination a achievable position solicitation of mind regions in which those locale turned out as most farsighted that are normally seen as essential for mental loaded with inclination torment taking care of for material getting ready. Considering alternatives among exactness (of the estimates) along with the multifaceted nature (of the model), a little course of action of anatomical locale of concern gave improved explanation of dynamic decision concerning torment than being voxels or else character zones [22–25].

The information-gaining procedure during N_c-channel d-magnetic resonance imaging with N_{fr} frame is capable of being formulated like

$$y_{ij} = S_i F C_j x_i + n_{ij}, \tag{14.1}$$

where S_i is the under-inspected design in ith outline, $i = 1,2,\ldots N_{fr}$, F is the Fourier framework, C_j is the affectability of the jth loop, $j = 1,2,\ldots N_c$, and η is the clamour. Characterizing the dynamic picture arrangement as a $N \times N_{fr}$ Casorati framework $X = [x_1, x_2, \ldots N_{fr}]$, where $N =$ number of stage encoding lines, $(N_p) \times$ Number of recurrence encoding lines (N_f), and every segment xi is the vectorized picture, (14.1) can be modified as

$$Y = \varphi(X) + \eta; \tag{14.2}$$

where φ is estimation administrator that consolidates Fourier under testing and loop affectability in (14.2). To remake dynamic picture arrangement X from obtained under-inspected k-space information, Y, an advancement issue, regularized by means of $R(\bullet)$ is fathomed:

$$\arg \ \min_X \|Y - \varphi(X)\|^2_F + \lambda R(X) : \tag{14.3}$$

Regularly, $R(\bullet)$ is the transient Fourier sparsity-mindful misfortune Ft(X) 1, where Ft($\bullet$) is the Fourier change along fleeting course or the low-position earlier rank(X). Nonetheless, this paper proposes two novel regularizers dependent on a complex-based methodology. As an initial step, neighbourhood connection among pictures is found out through our fundamental complex-based displaying. Furthermore, a nonlinear mapping which implants M into CM, $M \ll N$, is registered, and two regularizers are characterized dependent on the scholarly complex geometry. At last, we implement those regularizers on undersampled k-space information to reproduce the ideal dMR picture.

Critical exertions have been complete towards expanding a convincing method to manage help inside the conspicuous evidence of untimely occasions of dementia. Whereas dimensionality reduction along with feature extraction is played important role distinctive continuous CAD approaches designed for timely acknowledgment of dementia of image without decreasing its dimensionality. On behalf of instance, the masters planned a framework to facilitate and perceive patients through Parkinson's disease beginning NC with surveying known precedents by an estimation of affectability, downsized mental state examination, and distinction score. As well,Rocchi planned move towards with specific moves up in the direction of separate among patients resolved to have dementia along with strong controls by removing ROIs from helpful cerebrum pictures.

Basic attempt has been completed on the way to expand a convincing method towards managing help inside the conspicuous verification of untimely examples of dementia. Whereas dimensionality diminish and feature extraction be up 'til now critical, diverse continuous CAD approaches design for untimely acknowledgment of dementia contain be determined on top of picture request exclusive of decreasing its dimensionality. On behalf of the pattern, the masters planned a framework so as to perceive patients through Parkinson's sickness beginning NC with evaluated known models by an estimation of affectability and minimum mean square error (MMSE), along with disposition score [31, 32]. While glowing, Rocchi et al. planned advance through specific moves up near separate among patients resolved to have dementia along with strong control through removing ROIs beginning remedial cerebrum pictures. A pair of moves progresses during investigate in the direction of remove the enormous features of the game plan of information contain be used auto-encoder. Researchers endeavoured to isolate patients based on AD with noise individuals near propose a blend of the SAE within direct towards take in features removed since magnetic resonance imaging pictures along with the convolutional neural networks to facilitate towards the issues of picture course of action. The effort planned an AD gathering subject to a significant learning-based component depiction with the aim of have be made with stacked insufficient auto-encoders. SAEs have been subjected to discover a latent depiction of commitment through expelling features similar towards the volume of diminishing tissue as well as power starting magnetic resonance imaging and positron emission tomography, independently. Consequent with the works the planned technique revolves around adjusting anomalous state structure information for isolating among a couple of sorts of dementia in its place of now AD by isolating among dull along with grey issue voxels towards make far above the ground bore with point-by-point input information of SAE.

14.3 Methods

14.3.1 Sliced Inversed Regression

Designed for a regression difficulty, an all-purpose model is able to outline the connection connecting perception $y_{(n \times 1)}$ (a vector of n tests of conduct information or else psychological parameter inside this examination) and indicators $x_{(n \times 1)}$ (a lattice of f-magnetic resonance imaging information with p voxels designed for every watched test) like

$$y = f (\mathbf{x}\beta_1, \mathbf{x}\beta_2, \ldots \mathbf{x}\beta_k, \varepsilon) \tag{14.4}$$

where $\beta_{(s)}$, $s = 1, \ldots, k$, is a set of obscure vectors signifying the commitment of every indicator inside x, k ($k < p$) be the measurement we expect towards decrease x towards, f (.) be an obscure capacity of k information sources, with ε signifies zero-mean irregular clamour autonomous of x. The all-purpose supposition of this mode is the reaction y which just relies upon x by means of k straight blends of indicators:

$$y = \mathbf{x} \mid P_s x, S = \text{spam} (\beta_1, \ldots, \beta_2) \tag{14.5}$$

where P signifies the bulge administrator onto the k-dimensional subspace S. In this manner, we just need to assess S created by β_s to viably lessen dimensionality.

The EDR space for (14.4) can be evaluated by the SIR strategy by means of decision of an opposite relapse bend $E(x|y)$. The point-by-point plot for SIR is given underneath:

1. Normalize $\mathbf{x}$ along with approximation of the sample mean $\overline{x} = \frac{1}{n}\sum_{i=1}^{n} x_i$ and the sample covariance matrix.

$$\hat{\sum}_x = \sum_{i=1}^{n} \frac{1}{n-1}(x_i - \overline{x})^T (x_i - \overline{x})$$

2. Bin y interested in m slices, $G_1, \ldots G_m$, furthermore computes the percentage of y_i that falls interested in slices G_j, $j = 1, \ldots m$, as $\hat{P}_j = \frac{1}{n}\sum_{i=1}^{n} \xi_j (y_i)$, where $\xi_j(y_i)$ equals 1 or else 0 depending on top of whether y_i fall interested in the jth slice or else not.
3. Designed for every segment, compute the segment mean $\hat{x}_j = \frac{1}{\hat{P}_j}\sum_{i \in G} x_i$ and weighted covariance

for the sliced means. Accomplish a biased PCA designed for $\hat{\Sigma}_x$ with $\hat{\Sigma}_W$ through solving a comprehensive eigendecompostion difficulty:

$$\hat{\sum}_W \lambda_1 \geq \lambda_2 \geq \cdots \geq \lambda_k \tag{14.6}$$

$$\hat{\beta}_s = \hat{\lambda}_s \hat{\Sigma}_x \hat{\beta}_s, \text{ where}$$

In Steps 2 and 3, we can acquire the appraisals of institutionalized opposite relapse bend $E(x \mid y)$. A notice to facilitate is required just before change cut G_j to x_j before PCA, instead of to change all xi. It was demonstrated that the quantity of cuts ordinarily would not fundamentally influence the yield gauges. For an order issue, it is likewise recommended to facilitate the quantity of cuts that exists equivalent towards the quantity of lessons.

14.3.2   Principal Component-Sliced Inverse Regression

Sliced Inverse Regression requires the quantity of perceptions n towards being more prominent than the quantity of indicators p, which vigorously constrains its appropriateness inside the examinations of high-dimensional neuroimaging information. When $n < p$, Σ x be particular, along with in this way the answer for the Eigen-disintegration isn't remarkable. To address this issue, we planned towards utilizing PCA towards lessening the quantity of indicators previous to SIR [20, 21, 31].

During PC-SIR, PCA original pre-forms institutionalized x towards discover l guideline parts (PCs), everywhere l equivalents towards the position of x. The connection of l PCs be indicated as $L_n \times l$ (of which every segment be a PC), with the covariance framework of $L_n \times l$, Σ L, be a full-position lattice. SIR is next performing on top of $L_n \times l$ to discover the EDR heading, along with activities l-dimensional PCs towards k-dimensional subspace. PC SIR technique deals with order & replace of $n < p$, which is connected adequately for the related indicators.

14.3.3   F-magnetic Resonance Imaging Pain Prediction

14.3.3.1   Experimental Design

The examinations of the information set accumulated from unclear 32 strong subject with recording parameter from in the diversion consider. We passed on 10 laser pulse by all of the four lift powers (E1: 2.5 J; E2: 3 J; E3: 3.5 J; E4: 4 J) by means of an unpredictable along with variable between redesign interval (ISI) some place in the scope of 27 along with 33 s. The overhaul powers were pseudorandomized. Subject be advised towards progress a slider towards charge the intensity of the horrifying impression evoked with the laser beat 15 towards 18 seconds behind every redesign, by an electronic optical straightforward scale since 0 (contrasting with "no anguish") to 10 (identifying with "torment as horrendous as it could be").

14.3.3.2 PC-SIR Analyses

To pre process the information, PCA is initially associated with expel full-position PCs beginning markers (entire personality f-magnetic resonance imaging records next to the fourth scope behind lift starting, anywhere cerebrum have greatest response). Thus, SIR will be performed on top of PCs towards discovering the EDR course close by the straight mixes of PCs by means of mainly farsighted ability towards Y. The vague parameter of SIR from inside the entertainment consider. Backslide burdens are obtain along with changed towards the main hole with proper the anguish perception along with SIR-construed incorporate set. Before separate the cerebrum territories are added to torment insight, a point-by-point one-precedent t-test next to zero by means of $p < 0.001$ uncorrected performed on top of the assessed backslide stacks transversely over subjects.

14.3.3.3 Pain Prediction

During this examination, we attempted the introduction of planned strategy by means of five-overlay cross-endorsement philosophy. Precisely, PC-SIR evaluate mind activation map (backslide loads designed for every voxel) from the readiness fundamentals along with gauges complete designed for the assessment primers with attractive the spot consequence of the cerebrum establishment maps along with complete cerebrum f-magnetic resonance imaging records, compliant a scalar foreseen misery regard. The desire displays will be moreover assessed with MAE: we differentiated the introduction of PC-SIR and SVR (support vector machine regression) along with PLSR. The conjecture of this technique shows a gander at using a solitary course repeated estimates examination of progress (ANOVA). The principal consequence may be tremendous, position hoc pair adroit t-test and are to be preformed.

14.3.4 Manifold Learning

Let xi, $I = 1, \ldots, N_{fr}$, be present in the $N \times 1$ vector type of size $N_p \times N_f$ pictures. We propose that xi lies on or near an M-dimensional ($M \ll N$) complex $M \subset CN$. The initial stage is towards discovering the area connection connecting $\{xi\}i = 1N_{fr}$. Subsequent to several learning applications, everywhere neighbourhoods be characterized with the Euclidean separation of the surrounding space CN, we rewrite on top of the efficiency of the obscure complex M, along with propose to facilitate every picture vector xi be capable of exist approximated through the relative mix of its adjacent picture vectors: $x_{ierr} = x_i - \sum_{n=1}^{Nfr} w_i^{nX_n}$ imitating surely understood properties of digression spaces of flat manifolds. Along these lines, each picture is identified with its neighbours with the weight vector wi. The weight vector is compelled to exist inadequate, i.e., every picture have a predetermined number of neighbours. Such region spaces are comprehend by accompanying 'l1-compelled least squares issue:

$$\omega^i \in \underset{\omega_i^H 1 N_{fr}=1, \omega_i^i=0}{\arg\min} \left\| x_i - \sum_{n=1}^{N_{fr}} \omega_i^n X_n \right\|^2 + \beta \left\| \omega^i \right\|_1, \tag{14.7}$$

where $1 N_{fr}$ is each of the single vector. The requirement $\omega_i^{H1Nfr} = 1$; $(\bullet)H$ means Hermitian transposition, discover to load whole positive to 1, evaluated relative neighbouring associations, $\omega_i^i = 0$ prohibits xi from being a neighbour of itself, $\beta \geq 0$ can be tuned to oblige the quantities of neighbours Kn Eq. (14.7) from the above equation is unravelled utilizing contentions like the minimization system dependent on top of the praised least square shrinkage and selection operator (LASSO).

It is essential to note down under Magnetic Resonance Imaging acquiring, complete pictures xi even though they are not open; they are the perfect yield. Generally low frequencies are examined in all housings, however sporadic under testing is replicated along stage training along with point even when showed up during, by virtue of Cartesian under testing. Such repeat signal, called "guides," are stacked facilitate describe neighbourhoods. We exhibit that such pilot lines are adequately satisfactory to depict neighbourhoods and assessed complex embeddings. From now on, guides $\hat{x}$ are used to assess loads using the above equation (14.7).

14.3.5 Manifold Embedding and Regularization

In accelerated d-magnetic resonance imaging, the spatial information is profoundly undersampled; anyway the fleeting heading is regularly completely tested. This permits to gain proficiency with the fleeting premise of the ideal unique picture arrangement. Commonly, the worldly premises are found out through the PCA or solitary esteem disintegration (SVD) of a pilot signal. In this paper, we abuse the area connection to adapt such fleeting premises. Give W a chance to be a $N_{fr} \times N_{fr}$ weight framework with sections. It is observed for W, the nonlinear inserting of the equation Ψ, so as to map xi $\in$ CN into CM is given by understanding [6–10]

$$\Psi \in \underset{\Psi 1_M}{\overset{\arg\min}{C^{M \times N_{fr}, \Psi \Psi^H = I_M}}}, \left\| \psi_i - \sum_{n=1}^{N_{fr}} \omega_i^n \psi_n \right\| \tag{14.8}$$

where $\Psi = [\psi_1 \, \psi_2 \ldots \psi_{N_{fr}}]$. The product $\Psi \Psi^H = I_M$, where I is $M \times M$ personality framework, prohibits inconsequential every one of the zero arrangements, though $\Psi 1_M = 0$ focuses the sections of Ψ on 0. The arrangement of (14.7) is given by the Eigen deterioration of proper $N_{fr} \times N_{fr}$ matrices. Describe matrix

$$K := (I - W)(I - W)^{\mathrm{H}}; \ L := D - W;$$

where I is the personality framework and D is a corner to corner network with nonzero entry $d_{\mathrm{ii}} := \sum_{n=1}^{N_{\mathrm{fr}}} \omega_i^n$. Based on these two matrices, we develop two regularizers $R_1\,(\cdot)$ and $R_2\,(\cdot)$ specific to M-magnetic resonance imaging R_1 and M-magnetic resonance imaging R_2, respectively, to reconstruct dynamic magnetic resonance imaging from undersampled k-space information.

14.3.5.1   M-Magnetic Resonance ImagingR_1 (Affine Combination)

The initial regularizer depends on top of our display to facilitate every picture (point on top of a complex) and is intentionly approximated to blend of its neighbours. Characterize the blunder network task $X_{\mathrm{err}} = X\,(I - W)$. Rudimentary variable-based math uncovers that the ith section $x_{i\,\mathrm{err}} = x_i - \sum_{(n=1)}^{N_{\mathrm{fr}}} w_i^n X_n$ gives the mistake of estimation to xi. In that capacity, to authorize the relative blend displaying presumption, the accompanying misfortune capacity is presented

$$\begin{aligned} \mathcal{R}_1(X) &= \|X\,(I - W)\|_F^2 = \mathrm{tr}\left(X\,(I - W)\,(X\,(I - W))^H\right) \\ &= \mathrm{tr}\left(XKX^H\right) \end{aligned} \tag{14.9}$$

where tr signifies the hint of a framework.

14.3.5.2   M-Magnetic Resonance Imaging$R2$

The Laplacian matrix L can be decayed as $L = GGH$, where GH represents the frequency lattice. It has been demonstrated that the grid G goes about as a limited distinction administrator and XG_F^2 takes after Tikhonov regularization. Consequently, characterizing the second regularizer as

$$R_2(X) = \|XG\|_F^2 = tr\left(XLX^H\right): \tag{14.10}$$

The regularizer $R_2\,(\bullet)$ resembles the "l2 STORM" and used along with the laplacian graph to match complex efficiency towards the arrangement. Regardless, unique, neighbourhoods which are customer portrayed and such information free-thinker, are described by methods for relative associations along with by means of discovering loads as seen in equation (14.7). Nonexclusive piece learn records associations may be direct along with to learning systems to facilitate use scattering maps and warmth stream, anyway such a "one-fits-all" move on normally has every one of the reserves of being slanted to exhibiting mistakes [12–19].

14.3.5.3   Manifold Regularization and Reconstruction

The solitary vectors of the $N_{\mathrm{fr}} \times N_{\mathrm{fr}}$ matrices K and L give the arrangements of
(14.8). These solitary vectors rough the worldly premise of dynamic pictures as
delineated within the allowed solitary esteem deteriorations of K with L

$$K = \Psi^k \Sigma^k \Psi^{k\mathrm{H}}; \tag{14.11}$$

$$L = \Psi^l \Sigma^l \Psi^{l\mathrm{H}}; \tag{14.12}$$

where superscripts $(\bullet)^k$ and $(\bullet)^l$ demonstrate the disintegration of K and L, separately.
Eigen deteriorations have been broadly used to speak to fleeting varieties in d-
magnetic resonance imaging. The merit of PCA disintegrations save information 20
covariances, the "decay" of (14.7) safeguards the complex geometry. Substituting,
(14.11) in (14.9)

$$R_1(X) = \mathrm{tr}\left(X\Psi^k \right) \Sigma^k \left(X\Psi^k \right)^{\mathrm{H}} = \mathrm{tr}\ U^k \Sigma^k \left(U^k \right)^{\mathrm{H}} \\ = \sum_{m=1}^{M} \sigma_m^k \left\| u_m^k \right\|^2 \tag{14.13}$$

where $U^k = X\Psi^k$. Not at all like in PCA, M minor particular vectors are utilized,
rather than vital, to rough the arrangement of (14.8). Consequently, $u_m^k = X\Psi^k$
will be the projection of X on top of the mth minor solitary vector of K. So also,
substituting (14.12) in (14.10), we acquire a comparable articulation for R_2:

$$R_2(X) = \sum_{m=1}^{M} \sigma_m^l \left\| u_m^l \right\|^2 : \tag{14.14}$$

Subsequently, the first issue in (14.6) can be communicated by means of U^k along
with U^i while

$$\arg\min_{U^k} \left\| Y - \phi\left(U^k \left(\Psi^k \right)^H \right) \right\|_F^2 + \lambda \sum_{m=1}^{M} \sigma_m^k \left\| u_m^k \right\|^2, \tag{14.15}$$

$$\arg\min_{U^l} \left\| Y - \phi\left(U^l \left(\Psi^l \right)^H \right) \right\|_F^2 + \lambda \sum_{m=1}^{M} \sigma_m^l \left\| u_m^l \right\|^2, \tag{14.16}$$

At long last, the conjugate angle calculation is utilized to tackle (14.15) and
(14.16) and recreate wanted pictures from U^k along with U^i.

These investigations contemplate orchestrate a couple of sorts of dementia with
low computational costs along with lower additional room. To objectives, SAE is
used on behalf of adjusting strange state features related among multi form dementia

commencing pixel forces of magnetic resonance imaging neuroimages. Raw Magnetic Resonance Imaging pictures data is preprocessed along with segmention, then it is changed SAE designed for evacuating the huge features. The expelled features yield by SAE can be used with ML classifiers to get an exactness figure of various sorts of dementia.

14.3.6   An Initial Image Processing

The unrefined magnetic resonance imaging pictures; secured which are of dissimilar size; but classifiers need reliable dimension of information vectors. These photos are pre-processed to match with the structure and executing strategy. Every magnetic resonance image is resized to identifiable dimension. 176 * 208 * 1, which results to 36,608 voxels. They are preprocessed by picture preprocessing methodology, along with mechanized spatial isolating frameworks and picture change technique the photos selected for neuroimaging strategies generally have commotion. All in all, the preprocessing step expects a gigantic activity in extending the resolute nature of the visual appearance of pictures and improving the photos' quality [13].

The standardization, Gaussian channel, and histogram equalization are the strategies in picture pre-processing to facilitate association with primary MAGNETIC RESONANCE IMAGING pictures designed for setting them up designed for CAD structure. At first, these magnetic resonance imaging pictures are institutionalized towards the practically identical extent of dark extent as of 0–1 with use of the condition. Due to the institutionalization the extent of power is decreased so as to streamline features considered for undertaking.

$$y = \frac{x - v_{min}}{v_{max} - v_{min}}$$

The Gaussian filter is used to expel the commotion in picture signal. Which is designed to getting better the pictures' to accomplish the objective through optimization.

14.3.7   Image Segmentation

Segmentation moreover has a critical activity in dismembering therapeutic photos of the cerebrum and allows versions used for the superior finding of dementia. Picture segmentation isolates the picture to isolate over range of area to keep objects of equivalent uniqueness seperately. In this investigation, two unmistakable division methodologies are associated on magnetic resonance imaging photos of the cerebrum to envision anatomical structures of the psyche and after that separate changes of that mind:

- Edge-based division: Detection of edges in the picture is one of the critical steps to recuperate the perfect in turn from given pictures. Edges are perceived by predicting the modifications in some various properties of the image like brightness by using any of edge markers. In this paper, the Canny edge acknowledgment is performed to perceive edges that address one of the basic features for looking at cerebrum pictures and shield confined edges from continuing.
- Region-based division: A picture could moreover be separated into areas. Each clustered area is incorporated with tons of adjacent pixels that share traits and the proportional greyscale in this regard. For applying this kind of division in this examination, the watershed change computation is used to convey better division results.
- Sparse auto-encoder method to predict dementia features.

AE is associated lately as a preprocessed adventure with advancement for various PC vision errands. AE is a phony neural framework (ANN) that can be finished for repeating input information to fit in lower-dimensional space. The structure of AE is made of many layers including an information layer, a yield layer, and one hidden layer. At first, the information layer addresses encoding limit $f(x)$, mapping the information set, $x \in \mathbb{R}n_{input}$ to compacted information $h \in \mathbb{R}n$. By then, the disguised layer that addresses a character work gauges the proportion of information setback between the pressed depiction of your information and the decompressed depiction; in conclusion, h is decoded to get $x^* \in \mathbb{R}n_{input}$ that is a reproduced depiction of the commitment by applying deciphering limit $g(x)$ at the yield layer. Regardless, the information and yield layers have a comparable number of neurons.

In this correction, input xi speaks to the preprocessed pixel forces removed from magnetic resonance imaging cerebrum pictures where xi = {x1, x2, x36608}. Encoder work with input x to compacted structure $h \in \mathbb{R}!$ that fits in constrained spaces u, which u depicts various shrouded units as in equation (14.17).

$$h = f(Wx + b) = W'x + b' \qquad (14.17)$$

where $W \in \mathbb{R}u \times n_{input}$ and $b \in \mathbb{R}u$ speak to the framework of weight and the vector of inclinations of encoder work, individually. Likewise, f is a sigmoid capacity.

At last, decoder capacity reproduces the information x by taking shrouded portrayal h as contribution towards obtain x to facilitate significant highlights designed for dementia order undertaking. Condition (14.18) is utilized towards concern decoder work.

$$x* = \left(W'h + b'\right) = W''h + b'' \qquad (14.18)$$

$g(x)$ works as genuine esteemed contribution where $W' \in \mathbb{R}u \times n_{input}$ and $b' \in \mathbb{R}n_{input}$ the weight network along with the predisposition vector of decoder work.

To confine AE with diminishing numeral of concealed units, sparsity limitation pc specified by (14.19) is constrained on top of shrouded unit u construct them and make the greatest low-dimensional portrayal of information x. To accomplish this, the initiation of each one shrouded element, the middle value of to be near given sparsity parameter [10]

$$pc = 1 / (n_{input}) \sum_{(i=1)}^{(n_{input})} a_{uj}(x_i) \qquad (14.19)$$

At that point, sparsity imperative of each concealed unit is added to advancement article to limit a blunder among them along with the specified sparsity parameter. Used for preparing SAE, root mean square proliferation is planned towards recovering computational effectiveness through limiting mistakes along with charge.

14.3.8 Classification Methods

The novel depiction by using a classifier of features is evaluated on top of the endorsement put in records by use a classifier. For setting up the SAE, a significant 2D convolutional neural network is created, which takes magnetic resonance imaging channels as information. The nuances of setting up the created convolutional neural network will be recorded is. Used for redesigning the neural framework, on backside causing more computation time used for setting it up, to restrict cost and quicken association rate. By stochastic point ordinary, single of commonly use backside causing computations, is use designed for convolutional neural network getting ready through changing its hyper-parameters, together with loads, learning rate, along with pre demeanours. On behalf of relative investigation, LDA along with Logistic relapse comprise be utilized in favour of trying the planned methodology. The small dimensional picture because info, the yield of classifier is affectability, explicitness, along with precision of genuine forecast of five sorts of dementia.

14.4 Conclusion

In the present examination, another directed estimation diminishing methodology, PC-SIR will be planned on behalf of separating high-dimensional f-magnetic resonance imaging records. Along with f-magnetic resonance imaging torment desire grades are exhibited to facilitate PC-SIR on every basic level improved execution in (1) perceiving authorization plans starting the total personality f-magnetic resonance imaging records and (2) predict torment perception following estimation decline. Subsequently, PC-SIR will be a capable estimation decline strategy for looking at high-dimensional neuroimaging records; we displayed an original complex-

based construction used for dynamic picture diversion beginning under-inspected
k-space records. The original records focused strategy take in the complex form
getting ready information, called "pilots," and made two novel procedures inside
the intricate framework for changing unique magnetic resonance imaging from
extraordinarily under examined *k*-space information. The planned techniques will be
affirmed by mathematical spirit in addition to progressing within vivo records. Wide
endorsement of the planned methodologies on top of progressively educational
lists along with application is needed. It must be seen to facilitate the planned
procedures which don't depend upon the periodicity supposition, not at all like
the forefront PS-pitiful strategy. This pushes the examination and execution of the
planned structure in various previous d-magnetic resonance imaging applications,
for instance, talk, lungs, and liver imaging. The planned strategies in like manner
develop the system of abusing from the prior information from complex geometry
in addition to increase of new goals used for active MR picture diversion from
underexamined information. The chapter depicts another methodology along with
advances: institutionalization, picture pre-processing and division segmentation,
incorporate pondering as dimensionality abatement, and portrayal. This technique
associated SAE used for adjusting small depiction of records on the way to get
better request undertaking. The planned strategy and the top tier techniques were
attempted among the OASIS information set. Our system performs superior than
anything previously seen in top tier strategies. The other methods and PCA were
differentiated similarly as getting pictures and irregular state Biological marker.
It has be seen that it can be enhanced more than PCA by diminishing the
dimensionality of records. Future work growth of the amount of SAEs might bolster
the from execution, thus overhauling the gathering exactness.

References

1. K.A. Norman, S.M. Polyn, G.J. Detre, J.V. Haxby, Beyond mind-reading: Multi-voxel pattern analysis of fMRI data. Trends Cogn. Neurosci. **10**, 424–420 (2006)
2. B. Mwangi, T.S. Tian, J. Soares, A review of feature reduction techniques in neuroimaging. Neuroinformatics **12**, 229–244 (2014)
3. I. Jolliffe, Principal Component Analysis John Wiley & sons, Ltd (2nd Edition 2005) doi.org/10.1002/0470013192.bsa501
4. A. Krishnan, L.J. Williams, A.R. McIntosh, H. Abdi, Partial least squares methods for neuroimaging: A tutorial and review. NeuroImage **56**, 455–475 (2011)
5. H. Shen, L.B. Wang, Y.D. Liu, D.W. Hu, Discriminative analysis of resting-state functional connectivity patterns of schizophrenia using low dimensional embedding of fMRI. NeuroImage **49**, 3110–3121 (2014)
6. F.D. Martino, G. Valente, N. Staeren, J. Ashburner, R. Goebel, E. Formisano, Combining multivariate voxel selection and support vector machines for mapping and classification of fMRI spatial patterns. NeuroImage **43**, 44–58 (2008)
7. X.B. Gao, X.W. Wang, D.C. Tao, X.L. Li, Supervised gaussian process latent variable model for dimensionality reduction. IEEE Trans. Syst. Man Cybern. B **41**, 425–424 (2011)
8. K.C. Li, Sliced inverse regression for dimension reduction. J. Am. Statist. Assoc. **86**, 316–327 (1991)

9. K. Brodersen, K. Wiech, E. Lomakina, C. Lin, J. Buhmann, U. Bingel, M. Ploner, K.E. Stephan, I. Tracey, Decoding the perception of pain from fMRI using multivariate pattern analysis. NeuroImage **63**, 1162–1170 (2012)
10. An analytics Vidhya. A comprehensive guide to data exploration. https://www.analyticsvidhya.com/blog/2016/01/guide-dataexploration/, 2016 [Nov 20, 2016]
11. Stephen Gould, Tianshi Gao and Daphne Koller Region-Based Segmentation and Object Detection In Advances in Neural Information Processing Systems (NIPS), 2009
12. OASIS, What Is OASIS. Internet: http://www.oasis-brains.org [May 5, 2016]
13. N. Upadhyay, S. Bhargava, Removal of noise in medical imaging data using modified decision based adaptive weighted algorithm. Int. J. Comput. Appl. **84**(13), 34 (2013)
14. R.C. Gonzalez, R.E. Woods, Digital Image Processing, 3rd edn. (2008) Prentice-Hall, Inc. Upper Saddle River, NJ, USA
15. World Health Organization, 10 Facts on Dementia. Internet: http://www.who.int/features/factfiles/dementia/en/, 2015 [Apr. 1, 2016]
16. Rocchio, Jr., J. J. Relevance feedback in information retrieval. In Salton, G., editor, The SMART Retrieval System: Experiments in Automatic Document Processing, pp. 313–323. Prentice-Hall, Inc., Englewood Cli s, New Jersey (1971).
17. S. Liu, et al., Early diagnosis of alzheimer's disease with deep learning. In Biomedical Imaging (ISBI), 2014 IEEE 11th International Symposium on, Pages 1015–1018. IEEE, 2014
18. S. Bhargava, A. Somkuwar, Hybrid filters based denoising of medical images using adaptive wavelet thresholding algorithm. Int. J. Comput. Appl. **83**(3), 18 (2013)
19. C. Wachinger, P. Golland, M. Reuter, BrainPrint: Identifying subjects by their brain, in *Proc Intl Conf Med Image Compute Comp Ass Intervent*, Lecture Notes in Computer, vol. 8675, (2014), pp. 59–70
20. D. Davatzikos, S. Resnick, X. Wu, P. Parmpi, C. Clark, Individual patient diagnosis of AD and FTD via High- dimensional pattern classification of MRI. J. Neuroimage **41**, 1220–1227 (2008)
21. A. Payan, G. Montana, Predicting Alzheimer's disease: A neuroimaging study with 3D convolutional neural networks, arXiv: 1502.02506, 2015
22. T. Bodea, *Segmentation* (Routledge, London, 2013)
23. Z.-P. Liang, P.C. Lauterbur, An efficient method for dynamic magnetic resonance imaging. IEEE Trans. Med. Imaging **13**(4), 677–686 (1994)
24. J. Tsao, S. Kozerke, MRI temporal acceleration techniques. J. Magn. Reson. Imaging **36**(3), 543–560 (2012)
25. M. Lustig, D. Donoho, J.M. Pauly, Sparse MRI: The application of compressed sensing for rapid MR imaging. Magn. Reson. Med. **58**(6), 1182–1195 (2007)
26. R. Otazo, D. Kim, L. Axel, D.K. Sodickson, Combination of compressed sensing and parallel imaging for highly accelerated first pass cardiac perfusion MRI. Magn. Reson. Med. **64**(3), 767–776 (2010)
27. D. Liang, E.V. DiBella, R.-R. Chen, L. Ying, k-t ISD: Dynamic cardiac mr imaging using compressed sensing with iterative support detection. Magn. Reson. Med. **68**(1), 41–53 (2012)
28. Z.-P. Liang, Spatiotemporal imaging with partially separable functions, in *Biomedical Imaging (ISBI), IEEE International Symposium on*, (2007), pp. 988–991
29. B. Zhao, J.P. Haldar, A.G. Christodoulou, Z.-P. Liang, Image reconstruction from highly undersampled (k,t)-space data with joint partial separability and sparsity constraints. IEEE Trans. Med. Imaging **31**, 1809–1820 (2012). [30] S. Lingala, Y. Hu, E. Dibella, M. Jacob, Accelerated dynamic MRI exploiting sparsity and low-rank structure: k-t SLR, IEEE Trans. Med. Imag. **30**, 1042–1054 (2011)
30. S. Lingala, Y. Hu, E. Dibella, M. Jacob, Accelerated dynamic MRI exploiting sparsity and low-rank structure: k-t SLR, IEEE Trans. Med. Imag. **30**, 1042–1054 (2011)
31. S. Bhargava, A. Somkuwar, Estimation of noise removal techniques in medical imaging data–a review. J. Med. Imag. Health Inform. **6**, 1–10 (2016)
32. S. Bhargava, A. Somkuwar, Evaluation of noise exclusion of medical images using hybridization of particle swarm optimization and bivariate shrinkage methods. Int. J. Electr. Comput. Eng. (IJECE) **5**(3), 421–428 (2015). ISSN: 2088-8708

Chapter 15
Classification of Biometrics and Implementation Strategies

Snehlata Barde and Ayush Agrawal

15.1 Introduction

The word Biometrics is taken from Greek words bio & metrics; the word bio indicates life, whereas metrics means to measure. The automatic recognition process for identification is based on two characteristics: physiological and behavioral traits of a person. Further we can define biometrics as the following:

- Biometrics refers to technologies that are accustomed to discover and acknowledge human physical characteristics.
- In the IT world, biometrics is commonly similar with "biometric authentication," a kind of security authorization supported by biometrics.
- Biometrics could be a technological and scientific authentication technique supported by biology and utilized in data assurance (IA).

Biometrics is an important application of digital image processing that is used for statistical analysis of biological data. It is technical term for body measurement and calculation. The biometrics provides great solution to security technologies. It deals with automatic recognition of a person using distinguished modalities which are also known as biometric traits [1–6].

S. Barde (✉)
Department of MATS School of Information Technology, MATS University, Raipur, Chhattisgarh, India
e-mail: drsnehlata@matsuniversity.ac.in

A. Agrawal
Computer Science and Engineering, Bhilai Institute of Technology, Durg, Chhattisgarh, India

© Springer Nature Switzerland AG 2019
G.R. Sinha (ed.), *Advances in Biometrics*,
https://doi.org/10.1007/978-3-030-30436-2_15

15.2 Traits

The biometric modalities are also referred as "measurement of the human body." Biometric technology can be categorized as physiological and behavioral modalities.

15.2.1 Physiological Biometric Traits

There are some modalities which are correlated to physical formation or the property of human structure [1, 7–12]. Physical characteristics of human structure such as the face, ear shapes, iris, fingerprints, thumb print, hand geometry, footprints, etc. are discussed in brief here:

Face Biometrics Face recognition technique is a process of recognizing a person through its features extracted from the face. This is a computer-based application for automatic identification or verification using captured video frames and images. This type of biometric system is a most commonly used technique. Face recognition methods used various types of work on facial metrics and eigenimages. Facial metric method depends on the particular features of the face such as position of different traits and also finds distance among them, whereas the approach of eigenface relies on differentiating faces based on its degree in fixed set values between 100 and 150 eigenimages (Fig. 15.1).

Ear Biometrics Ear biometrics uses the ear as a modality where features or characteristics of the ear are used as the basis of matching. This is a firm biometric system which does not change with age. The ear is an observable part of the body that worked for a noninvasive procedure. The ears undergo a very little change in different stages of life. The ears cannot alter their appearance through hair growth similar to the face (Fig. 15.2).

Fig. 15.1 Face

Fig. 15.2 Ear

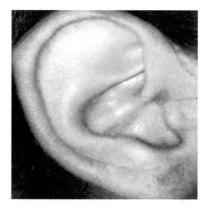

Fig. 15.3 Footprint

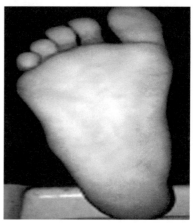

Footprint Biometrics The process of measuring features of footprint for recognizing a human. This system doesn't need any special device for image attainment. The image of a left/right leg can be captured from different angles. No light effect is used in the system. Through the key points, the images can be cropped and positioned. There are many different techniques used for resizing images to attain features. In database, distance technique has been used for comparing the feature vectors (Fig. 15.3).

Fingerprint Biometrics A fingerprint means ridge impression and valleys of finger and thumb images. The raised portion of the palm is known as friction ridge or toe of finger it is connected units of friction ridge skin. Generally ink is used to get the fingerprint on a paper, but nowadays, fingerprint scanners are used for detection. Optical fingerprint reader is the most commonly used which stores the reflection of fingerprint when finger lines stroke on the surface. This biometrics has some difficulties when the finger is dirty or wet (Fig. 15.4).

Fig. 15.4 Fingerprint

Fig. 15.5 Iris

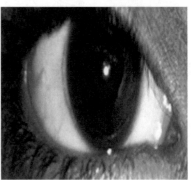

Iris Biometrics Colored and circular area of the eye is used for iris identification that isometrics background the pupil. Every person has unique iris patterns. A suitable method is used for image acquisition. Although the biometrics is not very user-friendly, it gives optimal performance (Fig. 15.5).

15.2.2 Behavior Biometric Traits

The traits that are related with behavior of persons are known as behavior characteristics such as signature, voice, keystroke, DNA, etc.

Voice Biometrics Voice of every person has different pitch, and hence it is considered as behavioral trait of a person. The voice recognition principally supported the voice study whenever someone speaks. Voice recognition is also known as speech recognition which is based on the characteristics of voice that produces speech and does not depend on pronunciation or sound. This biometric system does not need any special device for voice acquisition (Fig. 15.6).

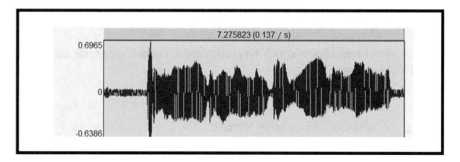

Fig. 15.6 Voice

Fig. 15.7 Signature

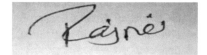

Fig. 15.8 Keystroke biometrics

Signature Biometrics Signature is a behavioral trait of a person based on the way person signs. Signatures are considered as the direction of writing, acceleration, pressure, and the length of dynamic strokes. The most important advantage of signature biometrics is that an impostor cannot collect the data on the way that it has been earlier written. Different types of devices are available for capturing the signature dynamics such as digital signature device and tablets (Fig. 15.7).

Keystroke Biometrics The identity of a person verified by this method is known as keystroke biometrics which is based on the way a person types and uses keystrokes on keyboard. The typing rhythm can be captured and helps in identifying a trained typists as well as an amateur two-finger typist with this type of biometrics (Fig. 15.8).

15.3 Classification of Biometric System

We classified biometrics into two types: unimodal biometric systems and multi-modal biometric systems.

15.3.1 Unimodal Biometric System

Unimodal biometric systems work on single traits of the individual for identification and verification [13–19].

Scope of Unimodal Biometric System
- *Commercial*: computer network log-in, e-commerce, ATM/credit cards, etc.
- *Government*: driver's license, passports, national ID cards, etc.
- *Forensic*: terrorist identification, identifying missing children, criminal investigation, etc.

Issues of Unimodal Biometric Systems These are the challenges of unimodal biometric system.

- *Noise*: A biometric data captured through an image acquisition system or sensor may be influenced by noise signal added due to sensor itself. This maybe due to imperfect acquisition conditions. Other factors that could contribute noise are subtle variations in the biometrics itself such that fingerprint image cannot be similar to the database due to scar or when the cold voice sample is altered.
- *Intra-class variations*: User does not properly know about the sensor device and how it works. So lack of knowledge on how to interact with sensor creates incorrect samples such as facial pose.
- *Non-universality*: The unimodal biometrics may not be providing accurate data from images of the database subset which store captured images. When the ridge quality of the finger is not good, it means that it does not display properly and then the features extracted from them will give incorrect fingerprint dataset.
- *Spoof attacks*: Human physiological trait such as the finger and behavioral trait such as voice or signature, are generally vulnerable to spoof attacks.

Drawbacks of Unimodal Biometric System The biometric system using single trait or characteristic are simple to use. However, these biometric systems have the following major drawbacks:

- To be short of universality.
- Noisy signals captured through the sensors due to the incorrect usage and due to the conditions of the environment.
- The discrimination of biometrics due to a high in-class and low interclass variability.
- The identification process of unimodal biometrics is restricted to a certain level.
- Error rates are sometimes unacceptable.

- The instability and changeability in biometric characteristics.
- The possibility of fraud through voluntary or involuntary cloning.

If there are problems with the trait being used, no alternative could save the biometric system.

15.3.2 Multimodal Biometric System

Multimodal biometric systems are nothing but a combination of two or more physiological and behavioral biometric traits. Biometric information through multiple sources are combined for overcoming some of the limitations mentioned in the previous system.

Goals of Multimodal Biometric System
- To calculate impostor rejection rate
- To find out the acceptance rate of impostors
- To indicate the enroll rate of failure
- To remove susceptibility of mimics

Advantages of Multimodal Biometric System There are many advantages of multimodal biometric system:

- *Accurateness*: The combination of multiple modalities is used to identify an individual in multimodal biometric system that improves the accuracy rate.
- *Liveness detection*: The system takes data as multiple traits randomly from the end users to provide powerful liveness detection and protection from spoof attacker or hackers.
- *Security*: This system eliminates all possible conditions of spoof that improves the security.
- *Cost-effective*: This system maintains the high level security to reduce the chances of crime and spoof that increases the effectiveness of cost.
- *Universality*: Nature of multimodal biometric system is universal, although someone is not capable to supply a variety of biometrics because of incapacity, so the system will choose different kind of method for authentication.

15.4 Implementation Strategies

Different level of fusions is available in this system. The main aim of this is to find the most excellent collection of dataset and develop a suitable role that could merge resultant data at optimum level. These are categorized as:

Prior to Matching

Fusion schemes prior to matching are used to integrate the evidence before matching. Two levels of fusion sensor and feature are important fusion schemes.

- *Sensor level*: Raw/unrefined collection of data taken as different modalities through multiple sensor devices are used for fusion means new data can be generated which can be used to extract the features. In this level fusion maybe performed if the information or modalities are acquired from multiple well-suited devices or various attributes of biometrics are obtained employing the particular device. This stage used different sensors for combining the information and applied some preprocessing before registration of data.
- *Feature level*: In this level of fusion different modality features are consolidated from the features obtained from using suitable methods of feature extraction. If the features are structurally compatible, then the features can be combined, and this is done by multiple sources. This approach reduces dimensionality by using feature transformation or feature selection for the fused feature set.

After Matching

Fusion schemes after matching are worked on processed data. There are three types of fusion as follows:

- *Match score levels*: Matching score is defined as a unit of evaluation in which the match level is calculated between the test data image and trained template stored in database, based on their biometric feature vector. Different matchers are combined for obtaining the match score. The match scores are not homogeneous because of different matchers applied on different modalities, remove this dissimilarity and use normalization technique that maps the different range scores on to a same range score.
- *Decision level and rank level*: When the fusion is at decision level, then each matcher's outputs are combined either to acceptance or to rejection. Decisions are taken by biometric modalities in decision level fusion. Binary value is holed by decision level fusion. The strategy adopted for concatenation of biometric modalities depends on the level of fusion performed. Fusions at the feature level are often achieved by combining two compatible feature sets. The method of selection or reduction of feature can be utilized for the problem of dimension. It is relevant to the identification of system in which classifiers are associated with a rank and each listed individuality.

Sensor level fusion and feature level fusion schemes require the data acquired by different sensors as compatible and feature set thus obtained by different traits. At matching, score level fusion is the most preferred because it has sufficient information and can be easily combined and is accessible. Decision level and rank level fusions take decisions using distinct modalities shown in Fig. 15.9.

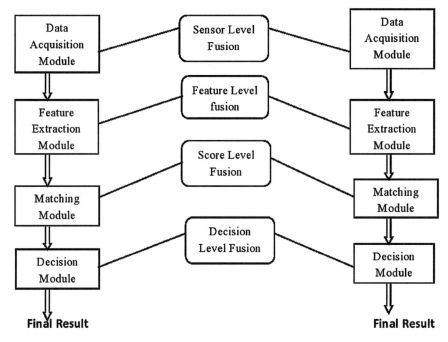

Fig. 15.9 Different fusion schemes

15.5 Set of Features

Biometric system provides result in terms of either a "genuine" or an "impostor" kind.

Genuine score Genuine score when it indicates result of an outcome of same user. A value that is less than the predefined threshold is genuine score.

Impostor score A score referred to as an impostor score when it indicates the result as an outcome of different users. A value that is greater than the predefined threshold is known as impostor score.

False accept rate (FAR) False accept rate means the chances of an impostor acceptance as a genuine individual. This indicates the number of people who is falsely accepted against enrolled number of person.

False reject rate (FRR) False reject rate means the chances of a genuine individual rejections as an impostor. This indicates the number of people who is falsely rejected with respect to the total number of enrolled people.

Equal error rate (EER) When accepted and rejected error rates are equal, it is called as EER in which comparison between two systems is needed. This is defined by the ROC where FAR and FRR assign the same value.

Relative operating characteristic (ROC) The ROC is plotted as a graph against the FAR and FRR.

Weight of biometric traits A technique used for fusion assigned weight to each biometric trait. For the weight for ith trait, W_i is calculated as:

$$W_i = \frac{1}{\text{EER}_i}$$

Normalized score The match score of the individual trait may not be homogeneous, and the match data of the output of different traits has dissimilar statistical distribution. Therefore, min-max normalization technique is used to calculate normalized score of each trait. The weight of particular trait of all biometric traits is calculated as normalized score:

$$W_i = \frac{\frac{1}{\text{EER}_i}}{\sum_{j=1}^{n} \frac{1}{\text{EER}_j}}$$

where equal error rate EER_j is for jth trait and the traits number are n.

Score after fusion (matching score) The sum rule based on fusion technique is used in the work. The score after fusion is calculated as:

$$S = \sum_{j=1}^{n} \left(W_j S_j \right)$$

where S_j is match score and W_j is the weight of jth trait, respectively.

15.6 Information Representation Through Features

We developed a multimodality-based biometric system that used face image, ear image, iris, and footprint modalities. The enrollment is a very important process involved in biometrics. This is illustrated in Fig. 15.10. The steps of biometric trait enrollment are:

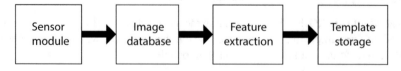

Fig. 15.10 Enrollment process in biometrics

- The biometric data or input is captured using suitable acquisition system or sensor.
- The modality is stored inside the biometric database.
- Features are extracted from the traits and converted into suitable transformations, called as biometric templates.

15.6.1 Training and Testing Involved in Biometric System

The two major processes, training and testing, divided into the biometric system. During training process, biometric modality is captured and converted into a suitable template. This process is performed as:

- Input image or signal is captured or acquired.
- The signal is preprocessed to remove noise or some similar signal; or image resizing or reformatting takes place.
- Feature extraction is performed.
- Features are transformed into suitable templates and these are stored in template database.

Testing process is performed at the time of matching, which is similar to training method. The input is captured, and it is subjected to:

- Preprocessing
- Feature extraction
- Template conversion

The template is matched against the templates already present in the template databases. If there is matching, it results in matching of the input. The training and testing processes are illustrated in Fig. 15.11, respectively.

All set of features were calculated independently and applied to different classifier approaches like PCA for face identification, eigenimage for the ear, Hamming distance method applied on iris identification, and modified sequential Haar transform used for footprint traits, respectively.

15.6.2 Principal Component Analysis for Face Recognition

PCA is the most popular algorithm used in face recognition. The main concept is to de-correlate the data which are eigenvectors of covariance matrix of a multidimensional data. Firstly, system is initialized with face image training set of vector. Testing of the biometric system makes use of face images from training dataset. Then, the trained images used PCA and trained dataset of images generated eigenvectors. The mean image value is computed as:

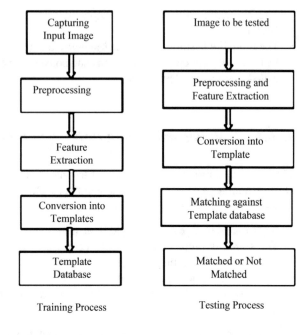

Fig. 15.11 Training and testing process involved in biometrics

Training Process   Testing Process

$$\psi_T = \frac{1}{N}\sum\nolimits_{n-1}^{N} \Gamma_n \tag{15.1}$$

where ψ_i is mean subtracted image.

The image ψ_i can be obtained by:

$$\psi_i = \Gamma_i - \psi_T \quad \text{where } i = 1, 2, \dots N \tag{15.2}$$

where vector set are subjected to PCA to get a set of N orthonormal vector set.

$$\lambda_K = \frac{1}{N}\sum_{n-1}^{N}\left(U_k^T \Phi_n\right)^2 \tag{15.3}$$

where eigenvectors are U_k and eigenvalues are λ_k, respectively.

The CM (covariance matrix) is given as:

$$CM = \frac{1}{M}\sum_{n-1}^{M}\left(\Phi_n \Phi_n^T\right) = AA^T \tag{15.4}$$

The Ψ is mean image computed, and through the eigenvectors, M is projected onto the "face space" resulting:

$$\omega_K = \mathrm{U}_K^\mathrm{T} \Phi_i \qquad K = 1 \dots M \tag{15.5}$$

Euclidean distance is used for calculation of distance between the training and testing projections as:

$$D_K = \| \Omega - \Omega_k \| \tag{15.6}$$

where kth face set is described by D_k.

Training set images are changed into face space and they are stored into memory. Face as an input is subjected and projected onto the face space. Then Euclidian distance is computed. If the image presented to the system is a face or not, it needs to be checked.

15.6.3 Eigenimage for Ear Recognition

Eigenimage method is the most effective method for ear recognition system. The ear recognition process is initialized by template images. The side pose of the face have been also acquired using high-quality camera in the same lighting condition. An ear portion is captured from the left/right face pose using preprocessing operation. The color images are converted to grayscale images which are subject for subsequent stages of biometrics. Each set of images contains images of training set and test set. In a huge dimensional space, an ear image can be considered as a vector with concatenating columns. The proposed method is based on normalized ear images that are preprocessed.

Ear image weights are stored by projecting the image onto an image space. Once the eigenspace is defined, the test image is projected into the eigenspace. The images with a less correlation may be rejected. Acceptance or rejection is determined by applying a threshold; less than threshold value indicates the match image.

15.6.4 Hamming Distance-Based Iris Recognition

The iris of an eye is the circular muscular structure that backgrounds pupil of the eyes. When light is incident, an eye is controlled by muscular structure called the iris. Iris system is supposed to be the most exact biometrics that utilize measurable features. Hamming distance method is used in iris identification in this work. Iris images are cropped from the eyes and then applied to preprocessing and encoding with Hough transform; the iris parts are localized to the eye image, and Hough transform-based segmentation algorithm is applied outside the pupil. The Hough

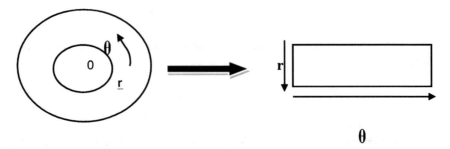

Fig. 15.12 The rubber sheet model

transform is a common method which is used in identifying features like location and orientations in a digital image. The technique is very simple and manages data which can be used in various forms not only lines. Iris recognition method involves three stages.

(a) *Segmentation*

Iris region is founded by eye image using segmentation process. It is approximated by two circles. The parameters are used (xc, yc) as center coordinates, and r is a radius. Edge defines threshold points, and both direction derivatives are detected by the outer area of the iris; for remapping, a rubber sheet model is used. This is shown in Fig. 15.12.

(b) *Normalization and Encoding*

The segmented area is changed to convert into dimensions. Due to pupil dilation, inconsistencies occurred. It has same constant after normalization of the regions of iris and also has constant dimensions. In this iris region, remapping of each points with (r, θ) which is coordinate of polar where $0 < r < 1$ and θ angle from $0 < \theta < 2\pi$ are represented by the homogenous rubber sheet model. During the matching, rotation is calculated by iris templates that are shifted in a different direction whenever they are aligned. Feature encoding extracted from iris region is achieved by one-dimensional Log-Gabor wavelets with pattern of the iris. One-dimensional signal is generated after dividing 2D normalized pattern. In 2D normalized pattern, 1D signal represented as rows, and these rows correspond to a circular region on the iris. The average intensity of surrounding pixels is normalized pattern of known noise areas.

(c) *Iris Matching*

Hamming distance is used for recognition as distance measure. It is defined between two templates used simply bits. These bits are corresponding to "0" that indicates noise masks of the iris, and this is modified by each template. The HD is defined as:

$$HD = \frac{\|C_A \otimes C_B \cap M_A \cap M_B\|}{\|M_A \cap M_B\|} \qquad (15.7)$$

where C_A and C_B are two iris coefficients of images and M_A and M_B are the mask image. Symbol $\otimes$ indicates XOR operator and $\cap$ indicates the AND operator.

15.6.5 Haar Wavelet Transform for Foot Recognition

Footprint is one of the unique modalities used for person identification to extract the features for calculation. Footprint of each person is different, and it does not change much across time and is easy to capture. Person's left/right leg is used for capturing the images of footprint to set in different angles without any special lighting arrangement. There are two basic transform methods discrete cosine and Fourier that are generally used for extracting the features of footprint, where the footprint image is cropped and positioned according to the key points.

Haar wavelet is used to take out the features of the foot. To measure the MHE is applied to the crop and resize image of footprint. Coefficients of Haar wavelet are described in the decimal numbers term. Feature is compared using Euclidean distance with stored feature vectors. Different decomposition levels are compared to find the accuracy. The samples of footprint of different people are cropped and resized. The images that are divided for obtaining MHE are calculated as:

$$MHE_{i,j,k} = \sum_{a=1}^{A} \sum_{b=1}^{B} (C_{a,b})^2 \qquad (15.8)$$

where i indicates the decomposition level; j denotes details of horizontal axis, vertical axis, or diagonal axis; 1 to 16 blocks are indicated by k; and the size is indicated by $A \times B$. The minimum Haar energy is selected out of 16 images. Let MHE_1 - MHE_{16} be the MHE values. Then a modified value is calculated by taking minimum of all the values. MHE = Minimum (MHE_1 to MHE_{16}). The MHE of different persons are compared.

15.7 Template Representation

We generated self-created database consisting of 200 person's images for the face, ear, iris, and footprint. Biometric information for every user were captured and stored as a training set. The necessary image preprocessing was also used so that images may be subjected to all other subsequent stages properly. After training process, the necessary features are extracted. Different classifier approaches at match score level for different modalities are used such as neural network-based

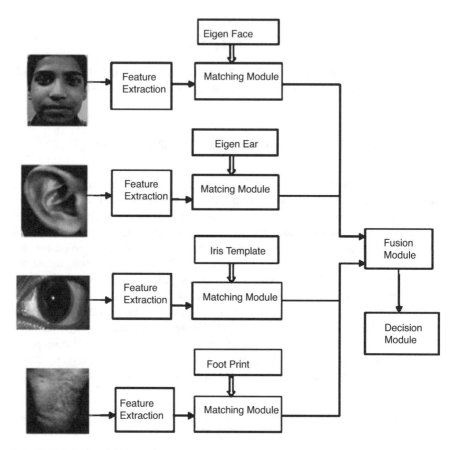

Fig. 15.13 Multimodal biometric system

principle component analysis for the face, eigenimage for ear image, Hamming distance-based approach for iris identification, and Haar transform technique for foot using MATLAB 7.10.0 software (Fig. 15.13).

15.7.1 Face Recognition

For face recognition, the front face images were acquired using high-quality camera in the same lighting condition with no illumination changes. The face image is saved using JPEG format. Preprocess of image included the cropped portion of the face from the image and then converted RGB image into grayscale image. The face images were resized into 170 × 190 pixels. Figure 15.14 shows few sample face images of the databases [20–23].

Fig. 15.14 Faces of the databases

Fig. 15.15 Training set

Figure 15.15 shows the training set of face images in PCA space corresponding to the transformation matrix. The individual images were normalized and then subjected to preprocessing operations, and clear face images were constructed. Figure 15.16 indicates a normalized dataset of face images.

Fig. 15.16 Normalized face images

Fig. 15.17 Face mean image

Figure 15.17 shows the mean image of the training dataset of faces; and Fig. 15.18 shows eigenfaces representing feature set.

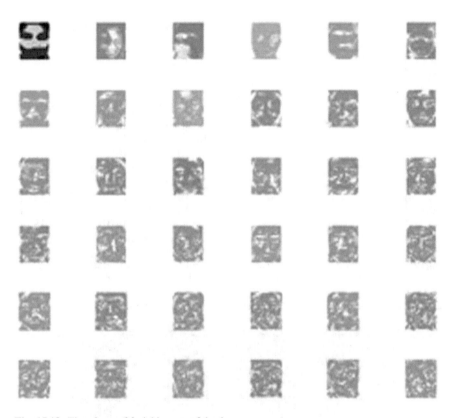

Fig. 15.18 Eigenfaces of facial images of database

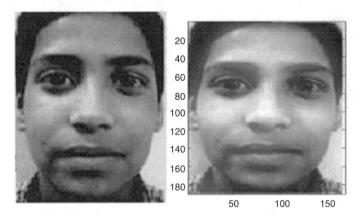

Fig. 15.19 Test image and its reconstruct image

Figure 15.19 shows test image as an input and output reconstructed image. Figure 15.20 shows weight and the distance of test face image. Euclidian distance helps to determine an acceptance or rejection of images.

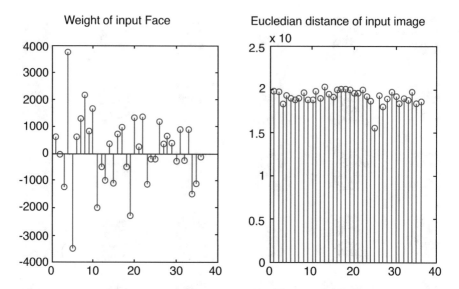

Fig. 15.20 Weight of face input and the Euclidian distance of input image

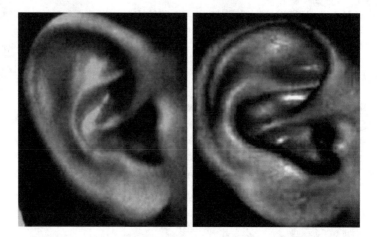

Fig. 15.21 Sample ear images of the database

15.7.2 Ear Recognition

For ear recognition, side face images were captured using high-quality camera in the same lighting condition with no illumination changes. Now, proper image preprocessing helped in getting cropped ear images. The ear images are saved in JPEG format. The RGB images are then converted to grayscale images and resized into 190 × 170 pixels. Figure 15.21 shows few sample ear images of the ear databases.

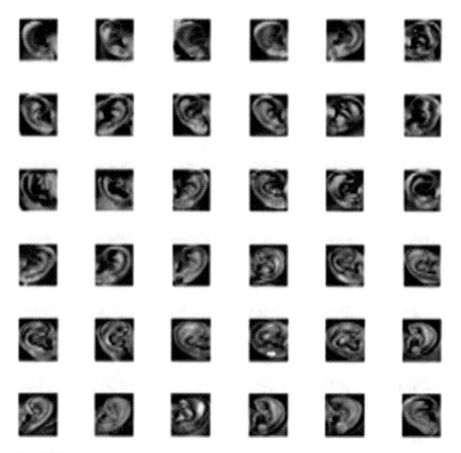

Fig. 15.22   Training set of ear images

In a similar manner, ear recognition was implemented. Training database of ear images shown in Fig. 15.22 and Normalized ear images represented in Fig. 15.23, eigenimage were estimated. Figure 15.24 shows the projection of images onto the image space, and stored their weights. Figure 15.25 shows result of biometrics tested over ear images.

Now, test image was projected into the eigenspace shown in Fig. 15.25. The threshold value determines the ratio of an accepted or rejected. When an unknown image was projected into eigenspace, then the distance from eigenspace to positions of unknown image is measured with respect to the position of all known image in eigenspace. The image which is nearest to the unidentified image in the eigenspace is found as matched image.

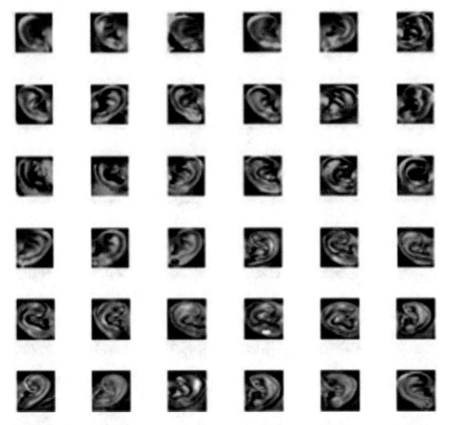

Fig. 15.23 Normalized ear images

15.7.3 Iris Recognition

In the human eye, the area lies between the cornea and the lens is called iris region. It is a thin circular diaphragm that has unique patterns and is not dependent to genetic factors. Two boundary circles can be used for designing of the iris region, sclera, and pupil. The eye images were captured using high-quality camera in a dark room with no lighting changes. The images are captured from a distance of 10–15 cm and saved in JPEG format. The eye part is cropped and converted to grayscale images and resized. Figure 15.26 shows sample iris images of the image database.

Iris recognition system is implemented and tested over databases of eye images. The system included segmentation, normalization, and feature encoding as important stages. Segmentation helped in locating the region of the iris in the image of eye as shown in Fig. 15.27.

Then the normalized iris region and extracted features help in producing templates as set of discriminated features of image. The eye image is used as an input and an iris template used as an output will provide a representation of iris region in mathematical. Figure 15.28 indicates the segmentation and normalization result.

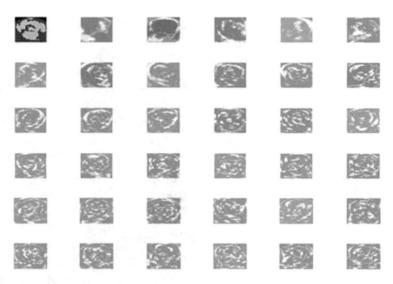

Fig. 15.24 Eigen ear images

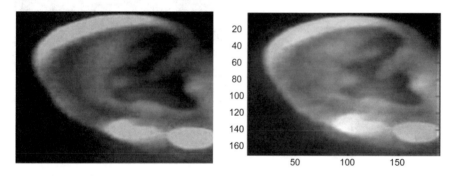

Fig. 15.25 An ear image and its reconstructed image

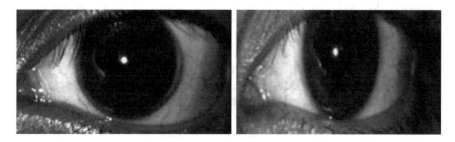

Fig. 15.26 Sample iris image of the database

Fig. 15.27 Segmented iris
image

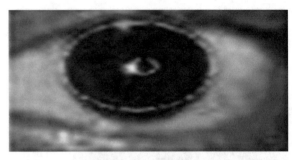

Fig. 15.28 Segmented and
normalized image

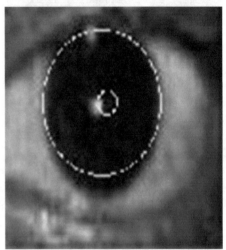

15.7.4   Footprint Recognition

For footprint recognition, footprints of hundred right leg images of different hundred
persons were captured with the help of digital camera without any special lighting
condition. The foot image was saved in JPEG format after acquiring the foot image,
key points were extracted in the image. The RGB format footprint image was
converted into grayscale image. Figure 15.29 shows samples of footprint images.

Now, foot modality was subjected to sequential modified Haar transform technique for foot recognition. It was mapped into integer-valued under the reconstruction property. The value of wavelet coefficients used decimal value storing in
the form of eight byte. For preserving the difference between two adjacent pixels,
Fig. 15.30 shows the result of foot image.

The middle portion of the leg was cropped because of its more intensity at this
portion; and the portion was arranged into 4x4 blocks. The comparison is performed
at different levels and the value of MHE is stored in database, and its minimum value
is selected. Finally, minimum MHE of test image is calculated.

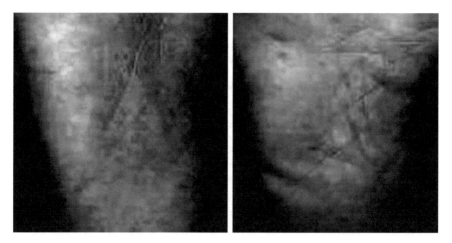

Fig. 15.29 Sample footprint images

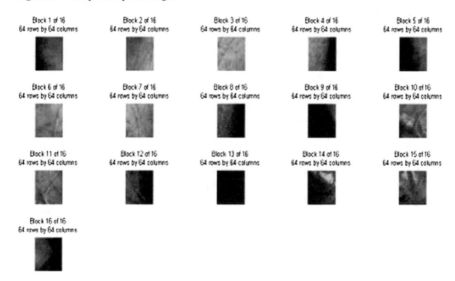

Fig. 15.30 Foot image divided into 4x4 blocks

Bibliography

1. A. Ross, A.K. Jain, et al., Information fusion in biometrics. Pattern Recogn. Lett. **24**(13), 2115–2125 (2003)
2. S.B. Kulkarni, U.P. Lulkarni, R.S. Hegadi, Analysis of iris recognition using normalized and un-normalized iris images. Int. J. Inf. Process. **7**(3), 26–33 (2013)
3. M. Turk, A. Pentland, et al., eigenfaces for recognition [J]. J. Cogn. Sci. **3**(1), 71–86 (1991)
4. D.R. Kisku, J.K. Singh, M. Tistarelli, et al., Multisensor biometric evidence fusion for person authentication using wavelet decomposition and monotonic decreasing graph [C], in *Proceedings of 7th International Conference on Advances in Pattern Recognition (ICAPR)* (Kolkata, India, 2009), pp. 205–208

5. W. Zhao, R. Chellappa, A. Rosenfeld, P.J. Phillips, et al., Face recognition [J]. ACM Comput. Surv. **35**(4), 399–458 (2003)
6. X.Y. Jing, Y.F. Yao, J.Y. Yang, M. Li, D. Zhang, et al., Face and palmprint pixel level fusion and kernel DCVRBF classifier for small sample biometric recognition [J]. Pattern Recogn. **40**(3), 3209–3224 (2007)
7. T.K. Ho, J.J. Hull, S.N. Srihari, et al., Decision combination in multiple classifier systems [J]. IEEE Trans. PAMI **16**(1), 66–75 (1994)
8. R. Cappelli, D. Maio, D. Maltoni et al., Combining fingerprint classifiers, in *First Internet Workshop on Multiple Classifier Systems* (2000), pp. 351–361
9. A.K. Jain, S. Prabhakar, L. Hong, et al., A multichannel approach to fingerprint classification [C]. IEEE Trans. PAMI **21**(4), 348–359 (1999)
10. A. Ross, R. Govindarajan, et al., Feature level fusion using hand and face biometrics [C], in *Proceedings of SPIE Conference on Biometric Technology for Human Identification*, vol. 5779 (Orlando, USA, 2004), pp. 196–204
11. J. Heo, S. Kong, B. Abidi, M. Abidi, et al., Fusion of Visible and Thermal Signatures with Eyeglass Removal for Robust Face Recognition, in *IEEE Workshop on Object Tracking and Classification beyond the Visible Spectrum in Conjunction* (Washington, DC, USA, 2004), pp. 94–99
12. G.R. Sinha, K. Thakur, et al., Modified PCA based noise reduction of CFA images [J]. J. Sci. Technol. Manag. **1**(2), 60–67 (2010)
13. L. Hong, A.K. Jain, et al., Integrating faces and fingerprints for personal identification [J]. IEEE Trans. Pattern Anal. Mach. Intell. **20**(12), 1295–1307 (1998)
14. R. Singh, M. Vatsa, A. Noore, et al., Integrated multilevel image fusion and match score fusion of visible and infrared face images for robust face recognition [J]. Pattern Recogn. ACM Digit. Libr. **41**(3), 880–893 (2008)
15. S. Singh, A. Gyaourova, G. Bebis, Pavlidies, et al. Infrared and visible image fusion for face recognition, in *SPIE Defense and Security Symposium*, vol. 41 (2004), pp. 585–596
16. A. Shrivastava, G.R. Sinha, et al., Biometric security technologies: a case study. Int. J. Image Process. Appl.. (International Science Press) **1**(1), 1–5 (2010)
17. W. Kong, D. Zhang, et al., Accurate iris segmentation based on novel reflection and eyelash detection model, in *Proceedings of 2001 International Symposium on Intelligent Multimedia, Video and Speech Processing*, (Hong Kong, 2001)
18. S. Barde, A.S. Zadgaonkar, G.R. Sinha, Multimodal biometrics using face, ear and iris modalities [c]. Int. J. Comput. Appl. Recent Adv. Inf. Technol. NCRAIT (2), 9–15 (2014)
19. S. Barde, Multimodal biometrics: most appropriate for person identification. i-manager's J. Pattern Recogn. **4**(3), 1–8 (2017)
20. S. Barde, A.S. Zadgaonkar, G.R. Sinha, PCA based multimodal biometrics using ear and face modalities. Int. J. Inf. Technol. Comput. Sci. **05**, 43–49 (2014)
21. S. Barde, A multimodal biometric system-aadhar card. i-manager's J. Image Process. **5**(2), 1–6 (2018)
22. S. Barde, Fusion based multimodal biometrics system. Int. J. Adv. Technol. Eng. Sci. **6**(5), 112–120 (2018)
23. S. Barde, Person identification using face, ear and foot modalities at rank level. i-manager's J. Comput. Sci. **2**l(2 1), 1–8 (2014)

Chapter 16
Advances in 3D Biometric Systems

Shankru Guggari and D. V. Rajeshwari Devi

16.1 Introduction

Biometrics is the asset of a person comprising of physical appearance and behavioral characteristics by which he/she is identified. In recent years biometric systems are successfully installed in many real-world applications like security, authorization, forensic science, etc. Biometrics efficiency is based on the location of data gathering, changes in the environment, and quality of individual interaction with the biometric system. Due to the effective learning capabilities, 3D biometric systems are gaining popularity in various real-world applications and scientific research. This chapter describes face, fingerprint, and iris recognition of 3D images which are proposed in the literature during the year 2009 to 2019. The chapter gives a generalized overview of face, fingerprint, and iris recognition of 3D biometrics and presents challenges along with their trends and prospects.

Some of the evolutionary algorithms like genetic algorithm, particle swarm optimization, and principal component analysis are used in the recognition of 3D biometric traits. Spoofing is a dangerous activity which causes serious effect to biometric systems and leads to abnormal detection. In this chapter, we discuss some of the anti-spoofing methods proposed for 3D biometric traits (face, iris, and fingerprint). Finally, a list of some popular open-source softwares for biometric identification is presented.

The reminder of the chapter is arranged as follows: a detailed advancement on 3D biometric systems are explained in Sect. 16.2; recent anti-spoofing methods are discussed in Sect. 16.3. A brief review of open-source softwares is given in Sect. 16.4, and the conclusions are given in Sect. 16.5.

S. Guggari (✉) · D. V. Rajeshwari Devi
B.M.S. College of Engineering, Bengaluru, India

© Springer Nature Switzerland AG 2019
G.R. Sinha (ed.), *Advances in Biometrics*,
https://doi.org/10.1007/978-3-030-30436-2_16

16.2 Developments in 3D Biometric Systems

Biometric authentication use physical or behavioral characteristics of an individual to identify him. Biometric systems work on two modes – enrolment (acquiring, processing, and storage of biometric samples into a database) and authentication (the test sample system is compared with the database samples to decide its genuineness). We discuss some of the recent major advancement techniques proposed for face, fingerprint, and iris recognition.

16.2.1 Face Recognition

Face recognition gain a significant attention in image analysis and understanding [1, 2]. Automatic recognition of facial expressions and moments are essential topics in computer vision. Facial image recognition attracts a lot of interest in today's world, because of more usage of digital image processing and computer graphics. There are some well-known challenges existing in the recognition of face:

- In real world, building a face variation model is a challenging task.
- Developing novel face detection techniques that are independent of:
  - Facial expression
  - Image condition
  - Pose of the face relative to camera face
  - Absence or presence of facial components such as mustaches, glasses, beards, etc.

- Developing hybrid face recognition system is a big challenging issue. It is the combination of both local features and holistic approach (enables quick recognition but not suitable for handling very large datasets).

A 3D face recognition technique is proposed to achieve robustness against facial expressions [3]. It uses elementary geometric descriptor, and superiority of the method is evaluated based on recognition rate and cross-validation recognition rate using GavabDB face and Notre Dame FRGC 3D databases [3]. A recognition technique is developed using the surface of the face and principal component analysis (PCA) [4]. It uses surface of the image, average curvatures, and Gaussian as input to PCA. A multilevel approximation technique using B-splines is used for facial surface normalization. Performance of the method is evaluated using ZJU3DFED database and achieves a recognition rate of 94.5% as compared to PCA [4]. In another study, a framework is introduced using simulated annealing (SA) approach for image recognition[5]. The Surface Interpenetration Measure (SIM) is used to match two images and gauge the similarity and get the authentication score combining SIM value and four different face regions (such as forehead, circular, full face region, and elliptical areas around the nose). It uses FRGC v2 database for experimentation which is composed of 3D face images (4,007 images) with various facial expressions and achieves 98.4% classification accuracy [5].

In recent years, face recognition using fuzzy logic is gaining popularity. A fuzzy rule-based matching technique is performed to recognize 3D face [6]. It uses Hausdorff distance to compute similarities among intra-class members. The superiority of the method is evaluated for both synthesized and original face images of Frav3D and GavabDB databases. It shows more than 7% improvement in classification accuracy as compared to original images [6]. The identification of scanned 3D face shape is performed by normalization [7]. This helps to detect the facial landmark and analyze the face shape based on the position of the 3D image. The method comprises of three important phases. Firstly, it converts 3D scanned image to 2D image and then extracts facial landmark features based on CNN and finally converts 2D image into 3D image. The classification accuracy is competent to other methods [7]. A 3D fuzzy GIST feature extraction is used for the analysis of EEG signals. It uses Support Vector Machine (SVM) classifier for classification and considers L^*C^*H color and information of orientation, based on the movie clip [8].

Feature selection is an important technique to improve the recognition rate. Recently an entropy-based technique is introduced to select the features to improve the facial expression classification rate [9]. It uses two-level SVM to avoid the confusions between the expressions. Experiments are conducted on U-3DFE database, and an average recognition rate of 88.28% is achieved, which is 8% higher than standard technique [9]. Feature selection is a best optimization factor in detection of human beings. A modified multi-objective method is introduced using genetic algorithm to improve the recognition rate in multimodal biometric system [10]. It uses incremental principal component analysis to take out the features and take support of genetic algorithm to achieve multiple goals by optimizing search space. It uses k-nearest neighbor classifier for classification and shows superior performance with respect to false acceptance ratio [10].

The evolutionary algorithms are well-known optimization algorithms which are widely used in various real-world applications. In designing biometric systems, some of the popular algorithms such as particle swarm optimization (PSO) [11] and Ant colony optimization techniques [12] are used to improve the recognition rate.

Fuzzy decision tree is used to classify the knuckle with training by Gaussian and trapezoidal membership functions and measurement of fuzzy information gain and Gini index. The optimal fusion parameters are chosen using Ant colony optimization technique with respect to the level of security [12]. More recently, a hybrid technique based on an evolutionary single Gabor kernel is proposed to detect the face [13]. It uses both particle swarm optimization and gravitational search algorithm to optimize the parameters in a single Gabor filter. It incorporates eigenvalue classifier to estimate the significance of the proposed technique as compared to other techniques like PCA and LDA [13].

16.2.2 Fingerprint Recognition

Fingerprint is a unique and highly reliable feature in human authentication. Traditionally, it is very popular in criminal investigation, and recently it is also used in applications like financial security and access control, etc. It suffers from some bottlenecks to achieve high recognition rate such as:

- Lack of novel feature extraction techniques.
- Reliable similarity measurement methods between fingerprints.
- Proper alignment of fingerprints.
- Identification of incomplete fingerprints.
- Lack of effective fingerprints matching method.

A partial or incomplete fingerprint identification is a challenging task. A region-based fingerprint method is proposed to overcome this challenge [14]. In this method pixel-wise technique is used to match the features with the help of correlation coefficient. It has 3 main important steps – alignment, extraction of common region, and computing the degree of similarity. The common regions are obtained by dividing the image into multiple smaller regions. The Fisher Transform is used to compute the local similarities and decrease the effect of distorted regions and use mean to find final degree of similarity between fingerprints [14]. More recently, minutiae points are used to detect partial fingerprints. It uses both bifurcation and termination minutiae points and a crossing number method (9,10, and 11) to scan pixels neighborhood [15].

A unique 3D fingerprint software is described using fringe projection [16] and uses patterns of color sinusoidal fringe. Experiments are carried out using three fringe numbers. The superiority of the method is evaluated based on standard deviation and absolute error [16]. The curvature features, like skeleton of the curve and overall maximum curvatures, are used for classification of human gender [17]. It uses 541 fingerprint database and shows promising equal error rate and also exhibits sectional maximum curvatures to do human gender classification [17]. There are various ways to construct 3D fingerprint. Recently, 3D fingerprint reconstruction is obtained using multi-view 2D images [18]. It reconstruct based on the correspondence with 2D images and use ridge feature, minutiae, and scale invariant feature transformation to establish correspondences. It undergoes hierarchical matching approach and proves that the binary quadratic function is reliable for shape of the finger [18].

In recent years, contactless 3D fingerprint technique attracts many researchers due to its ability in ubiquitous personal identification and accurate recognition [19]. It uses convolutional neural network (CNN) learning model with three siamese network and fully connected layers and take support of multi-view contactless dataset to understand the performance along with contactless 3D fingerprint database. ROC curve is drawn to indicate the novelty of the method and P-value for statistical significance [19]. A hand-based contactless biometric system is proposed using infrared imagery. During image collection, it captures images in infrared and RGB format and uses fuzzy-weighted technique at fusion level to access the image quality for multimodal biometric authentication [20].

16.2.3 Iris Recognition

In this era, authentication and security become two important and mandatory issues in the biometric systems. Iris image-based biometric systems have varied applications such as lend access to premises, tracing wanted and missing human beings, maintain render authentication in ATM, and obtain attendance report for very large-scale corporate systems. Although various iris biometric systems are available in the market, there are some open challenges existing as given [21]:

- Establishing benchmarks to make consistent in iris recognition without considering age and related issues with eye and its diseases.
- Develop novel techniques which counter balance the spoofing.
- Improper capture of the iris image with respect to symbol perspective.
- To explore iris recognition through deep learning techniques and artificial intelligence.
- Develop superior techniques to transform 2D iris images to 3D iris images.
- Techniques to detect fake iris image.

There are various methods to construct 3D iris model such as microplenoptic camera and Python Photogrammetry Toolbox. Age, dropping of eyelids movement, and segmentation of eyelash are the major problems in the detection of iris. A novel method is proposed to obtain 3D images of detached retina [22]. In this method, 12 partial retinal images are taken in clockwise direction and then cut into 12 sectors and resized to near relative sectors. The color information of 2D image is extracted from sphere mapping algorithm, and visualization tool kit is used to create 3D image [22]. A 3D iris image technique is proposed for the detection of iris [23]. Due to the poor lighting and eyelids and eyelashes movement, good quality iris images are not possible to obtain. A 3D iris image is constructed based on the salient fiducial points of two 2D images with the help of triangulation and random sampling consensus algorithm to map corner points [23]. A 3D eyeball tracking method is proposed with low response time and error. It estimates the eyeball movement and eye features during page scrolling [24].

A fuzzy-based classification is performed to detect both palm and iris [25]. The valley detection and neighbor pixel value algorithms are used to detect region of interest in palm and iris, respectively. It also uses statistical-local binary pattern technique to extract local features and bring the feature vector in the same range using Max normalization. Palm features are enhanced using both histogram of oriented gradient (HOG) and discrete cosine transform (DCT), and features of iris are extracted using Gabor-Zernike moment [25]. The Gabor filter is used to extract palmprint and iris features in four orientations with two-level wavelet decomposition [26].

In networking environment, biometric identification faces several challenges, such as leakage of users privacy and storing of users biometric template. Fuzzy vault and fuzzy commitment methods are used to store short keys and biometric templates. A crypto biometric scheme is used to retrieve a secret key based on iris template with the help of fuzzy extractor [27]. Fuzzy extractor is a biometric tool to authenticate the user based on his/her own template. It has two phases: one is enrollment phase where iris template masks the secret key, and another phase is called verification, where secret key is returned based on the similarity of both reference and query template. The Hamming distance is used to understand the variability of both intra- and inter-users. The efficiency of the method is evaluated on CASIA iris database [27]. In another study, fuzzy commitment technique is used to assign the secret key [28]. This technique randomly assigns a secret key with respect to the subject along with the binary features. Iris fuzzy commitment system is developed to improve both security and privacy. The Reed-Solomon and Hadamard codes are used to assess both security and privacy, and Markov chain model is used to describe the iris distribution. Experiments on CASIA iris database show the superiority of the method [28].

Some researchers have used evolutionary techniques to achieve better recognition rate for iris and obtain minimum number of features. Particle swarm optimization technique is used to get the features from iris and control it by using fuzzy rules. Haralick method is used to extract the crucial features from the iris [29]. In another study, iris features are extracted from Daubechies wavelets, and a subset of informative features are obtained from genetic algorithm. The WVU, UBIRIS Version 2, and ICE 2005 datasets are used to indicate its superiority. SVM is used to showcase the recognition rate [30].

In the next section, we discuss some of the recent anti-spoofing methods designed to overcome intrusions in 3D biometric systems.

16.3   Anti-spoofing

Spoofing is a technique to obstruct the normal operation of a biometric system in order to gain unauthenticated access. Anti-spoofing is a counter measure for spoofing and implemented using following approaches [31].

- Installation of additional hardware: It is costlier and can be invalidated using biometric traits.
- Accumulation of extra data (information) to distinguish the features.
- Use of authenticated biometric data which is captured from biometric systems.

The intrusion can happen at sensor level or feature level [32]. At sensor level, intruders pose as clients by presentation or direct attacks like, video, mask, and photo attacks. At feature level, the intruders try to modify the captured biometric data. The sensor-level spoofing is overcome by devising sensors or algorithms that distinguish fake and real faces. The feature-level spoofing is overcome by providing protection to biometric templates in the form of encryption techniques. In this study, we focus on sensor-level attacks.

Two levels of anti-spoofing methods are performed:

(i) Hardware: It uses specialized devices along with the sensor to detect character- istics of a live trait like blood pressure, facial thermogram, fingerprint sweat, reflection properties of the eye, etc.
(ii) Software: These are applied on the sensor data to extract features that distinguish a live and fake face. Further, software-based methods are classified as static and dynamic methods. The static methods are applied on still images like photographs, while dynamic methods are applied on video sequences. The photo and video attacks are categorized as 2D face recognition system, while mask attacks are categorized as both 2D and 3D face recognition systems. We discuss some of the methods proposed in the literature to overcome photo, video, and mask attacks.

In next section, we describe anti-spoofing methods for face, fingerprint, and iris recognition.

16.3.1   Face Anti-spoofing

The deep learning methods like convolutional neural network are incorporated for efficient anti-spoofing in face recognition system. A novel approach to anti-spoofing using noise modeling and denoising algorithms is proposed [33]. This method handles anti-spoofing for paper attacks and replay attacks. It performs face de- spoofing by decomposing it into noise pattern and live face. The measurement of the noise pattern is done using convolutional neural network. The degradation of the live face occurs in the given steps: color distortion, display artifacts, presenting artifacts, and imaging artifacts. This architecture consists of three parts: (i) the De-Spoof Net (DS Net) which estimates the noise pattern of the image and reconstructs the live face by subtracting input image from estimated noise (ii) the Discriminative Quality Net (DQ Net) and (iii) Visual Quality Net (VQ Net) which are used to control the visual appearance and liveliness of the reconstructed image. This method is tested on three face anti-spoofing datasets, Oulu-NPU, CASIA-MFSD, and Replay-Attack [34]. It uses metrics like Attack and Bona Fide Presentation Classification Error Rates and Half Total Error Rate to indicate the superiority.

A convolutional neural network (CNN) is used to learn the features of the face [35]. Anti-spoofing of face is performed using temporal, color, and patch- based features. Temporal features are converted into gray images and fed into CNN. For color features, the RGB image is transformed into HSV and YCbCr color spaces, since they are found to have more discriminative features and then fed to CNN to build the model. It creates equal size patches to extract the local information and used to train the CNN formed by 18-layer residual network. Each CNN outputs the probability whether a face image is live or fake. Finally, a SVM combines all the probabilities and classifies the given face image as live or fake image. Three databases are used for experimentation, CASIA-FASD [36], OULU-NPU, and REPLAY-MOBILE. This method is measured using equal error rate by incorporating three different face features.

A novel face anti-spoofing is introduced based on textural features and depth information of the face [37]. A CNN is used to train the texture features of face region and background. The depth images are captured from Kinect which is used along with the webcamera. The face regions of live and fake depth images are captured, and then the depth features are extracted using LBP. The video sequences are captured using camera and kinect. The final decision is based on decisions from CNN based texture features and Kinect-based depth features. The input image is classified as live if both the decisions classify it as live else it is classified as fake. A dataset containing depth information of 20 persons is generated using Kinect and RGB camera. This method exhibits lower Half Total Error Rate by combining both texture and depth information.

Long short-term memory (LSTM) units with CNN is implemented to deal with face spoof attacks [38]. The spatial features of video frames are extracted through deep neural network. Temporal features are fed to LSTM units for classification. The input to deep residual network, ResNet-50, is color image of size $3 \times 224 \times 224$. A 2048 feature vector is extracted from the CNN and fed to 256 LSTM units. Finally, classification is performed using softmax as the decision function. It uses CASIA-FASD [36] and Replay-Attack [34] databases to clarify the novelty of the method and achieve lesser error rates as compared to static and dynamic feature-based methods.

Some methods are proposed based on extracting color texture features and depth of the information. Color texture Markov feature extraction and redundant feature elimination using SVM are introduced to recognize the face [39]. Initially, adjacent pixels of face image are analyzed, and Markov process trains the model to classify real and fake images for each color channel. To capture the differences between adjacent pixels, directional difference filtering is employed. It is found that the consistency between adjacent pixels is deteriorated for a fake image in comparison to live image. Further, color channels are explored using mutual texture information. Finally, SVM-recursive feature elimination method is adopted to select distinct features based on weight magnitude, which is a ranking criteria. The SVM classifier gives the final decision for the input face image as a fake or live image. Experiments are conducted by using Oulu-NPU, CASIA-FASD [36], MSU-MSFD, and Replay-Attack databases [34] and show the superiority of the method over PCA-, LDA-, and LBP-based methods.

A new feature extraction method is introduced based on analysis of linear discriminant and legendre moments to extract the face features [40]. The maximum likelihood classifier calculates the Gaussian probability of the feature vector. This classifier is efficient when the variance around mean is narrow and the overlap between various classes is small. The likelihood of a real face belonging to a class is high in comparison to 3D mask face. The experiments conducted on 3DMAD database [41] achieve a recognition rate of 97.6% and spoof FAR of 0.83%. The proposed face recognition system also includes the task of verification, thus avoiding a separate verification stage.

A superior anti-spoofing technique is proposed for face recognition based on gradient texture information and weighted gradient-oriented feature vector from the depth map. Texture properties of the image enable to identify whether face is real or fake. Extensive experiments are performed to find the efficiency of the introduced method using Replay-Attack, NUAA imposter, and CASIA datasets and use detection rate as a metric [42]. Facial recognition systems are vulnerable to mask attacks. A 2D recognition system is proposed to deal with 3D mask attacks. The angular radial transformation (ART) method is used for feature extraction wherein images are projected orthogonally on a radial basis. It includes the features with both imaginary and real part for each circular moments. Further, feature reduction is achieved though LDA which enhances between class variations. For classification, nearest neighbor and maximum likelihood (ML) methods are used. The method (ART + ML) is tested on 3D Mask Attack database and is found to perform better than LBP + LDA method. Among the classifiers, the ML classifier exhibits lesser Half Total Error Rate [43].

A multimodal approach of feature-level fusion of different color space features is proposed for face spoof detection [44]. The RGB color space does not show any difference in terms of luminance and chrominance information and use HSV and YCbCr to extract color spaces. The face features in these color spaces are extracted using Enhanced Discrete Gaussian-Hermite moment-based Speeded-Up Robust Feature descriptor. Different band images are fused using Oppositional Gray Wolf Optimization algorithm by assigning optimal weight scores and used K-SVM classifier for classification. The classifier is a combination of k-NN classifier and multiple k-SVM classifiers connected serially to classify the image by using CASIA-FASD [36], Replay-Attack [34], and MSU-MSFD databases. In comparison to other classifiers like CNN + Backpropagation and CNN + Levenberg-Marquardt, the proposed classifier achieves higher recognition rates and lesser error rates.

A method to generate 3D face spoof data using virtual synthesis is proposed [45]. Since deep learning-based methods require extensive training samples, this method provides a solution by generating virtual data. A printed photo is transferred to 3D object, and its appearance is manipulated in 3D space. The 3D face object is meshed using Delaunay algorithm. After meshing, transformation operations like rotating and bending are applied on the 3D meshed face. To overcome the imbalance between spoof samples and live samples, (a) ratio of sampled live and spoof instances is fixed during training, or (b) external live samples are imported. This method is tested on CASIA-MFSD, CASIA-RFS, and Replay-Attack databases [34] and is found to perform better in terms of Attack and Bona Fide Presentation Classification Error Rate, Average Classification Error Rate (ACER), and Top-1 accuracy.

16.3.2   Fingerprint Anti-spoofing

A novel local descriptor is proposed for fingerprint liveness detection [46]. The Weber local binary descriptor computes the pixel variations in a local image patch by considering the background intensity also. In addition, the proposed descriptor extracts gradient orientation from center-symmetric pixel pairs. The feature vector is of size 944 with 8 neighborhood pixels. The Mahalanobis distance is used to compare the multivariable distributions and compute the recognition rate using SVM. The experiments are performed using LivDet2011DB, LivDet2013DB, and LivDet2015DB databases and exhibit lesser error rates. An approach utilizing multiple features (gradient and textural) for fingerprint liveness detection is proposed [47]. The low-level gradient features are extracted using Speeded-Up Robust Features (SURF) which is invariant to illumination, scale, and rotation. To overcome the variations due to geometric transformations, the local shape information is extracted using histogram of orientation gradient (PHOG). Also, texture features are captured using Gabor filters. For dimensionality reduction of SURF+PHOG and Gabor features, PCA is applied. The classification is performed using SVM and Random Forest, and dynamic score level combines the results. The experiments conducted on LivDet 2013 fingerprint database reveal that SVM performs better for SURF + PHOG features and Random Forest performs better for Gabor features.

A convolutional neural network approach is proposed for fingerprint liveness detection [48]. A comparison of four models is performed. The first model uses a convolutional neural network with random weights (CNN-Random) for feature extraction, followed by dimensionality reduction using PCA. Recognition rate is obtained by building model using SVM with RBF. The second and third models are CNN-AlexNet [49] and CNN-VGG [50], respectively, which are pre-trained for natural images. The fourth model extracts features using binary patterns which are present locally. The histogram image of LBP is further reduced by PCA and classified using SVM classifier. The experimentation is carried out on LivDet 2009, 2011, and 2013 databases. The CNN-VGG model exhibits the least error rate in comparison to other models.

16.3.3   Iris Anti-spoofing

Iris liveness recognition is based on quality assessment parameters [51]. It uses 22 features pertaining to focus, motion, occlusion, contrast, and dilation properties. Pixel intensity, angle information using directional filters, etc. are gathered from different sources. To overcome the issue of large dimensionality, Pudil's sequential floating feature selection algorithm is used for feature selection. Finally, iris image is classified either fake or real using standard quadratic classifier. From the experiments conducted on BioSec baseline database, it is found that individually the occlusion features exhibit least classification error rate, while the combination of all the features (occlusion, dilation, contrast, and others) reaches a zero error rate. A novel work based on Laplacian decomposition for iris images is proposed to overcome presentation attacks in visible spectrum and near-infrared iris systems [52]. It decomposes each image into multiple scales of Laplacian pyramids. At each scale, short-time Fourier transform (STFT) is applied to obtain responses in four directions (0, 45, 90, and 135 degrees). The histogram is formed for each response and generates a feature vector. The final vector is the concatenation of all the feature vectors of each scale in four directions. Finally, classification is performed using SVM with polynomial kernel. It uses presentation Attack Video Iris Database obtained from iPhone 5S and Nokia Lumia 1020 and exhibits a classification error rate of 0.64%. The proposed system is also efficient in achieving an error rate of 1.37% for LivDet iris database comprising of near-infrared images.

A methodology based on pupil dynamics for eye liveness detection is proposed [53]. The pupil dynamics is expressed in terms of change in its size and shape which is considered as a circular approximation. Hough transform is used to localize pupil in each frame, and a directional image representing the image gradient and direction is generated. Each iris image is converted into a time series of pupil radii. The gradient values above a certain threshold are only considered, and if not even a single gradient is above the threshold, the pupil is not detected. The dilation and constriction of pupil in the presence of variation in light intensity are modeled by Kohn and Clynes and transformed into a seven-dimensional feature space. Finally, SVM is used for classification using linear, radial, and polynomial kernels. A self-generated dataset is used to compute the performance, and acceptable error rates are obtained. However, this method has drawbacks, like the time required for capturing the pupil dynamics, variations in pupil with age, and psychological conditions.

16.4 Open-Source Softwares

In this section, we introduce some open sources which are proposed in the literature. An OpenBR Collaboratory is an open-source project for the development of biometric research having an introduction to 4SF face recognition algorithm [54]. The ImageWare Systems maintains an open-source project called Open Biometrics Initiative (OBI) [55] having two APIs – OpenEBTS API based on Electronic Biometric Transmission Specification standards and OpenM1 API based on the INCITS and ISO standards.

BioSecure NOE [56, 57] has developed many open-source systems using publicly available datasets. It has modalities for iris, fingerprint, hand, signature, speech, and talking face. The signature modality is based on Hidden Markov Model and Levenshtein distance and uses geometry of fingers to recognize hand modality with support of MCYT-100 benchmark dataset [58]. The BioSecure reference system developed two algorithms, closet iterative points and thin plate spline warping for 3D face modality using 3D RMA database. It also developed modality for iris using

CBS dataset [59]. The BioAPI is open source for biometric technology to provide single sign-on web authentication system [60].

16.5 Conclusions

The challenges of 3D biometric systems with respect to face, fingerprint, and iris are presented. The recent advancements in these systems are discussed. A detailed explanation about various anti-spoofing methods are discussed to overcome intrusions from impostors. Finally, an overview of existing open-source softwares is mentioned.

References

1. J. Sushma, B.S. Singh, R.S. Jadon, D.T. Kumar, Brief description of image based 3D face recognition methods. 3D Res. **1**(4), 1–2 (2011)
2. K.W. Bowyer, K.P. Hollingsworth, P.J. Flynn, A survey of iris biometrics research: 2008–2010 (2016), pp. 23–61
3. X. Li, H. Zhang, Adapting geometric attributes for expression-invariant 3D face recognition, in *IEEE International Conference on Shape Modeling and Applications 2007 (SMI'07)* (2007), pp. 21–32
4. L. Yunqi, C. Dongjie, Y. Meiling, L. Qingmin, S. Zhenxiang, 3D face recognition by surface classification image and PCA, in *2009 Second International Conference on Machine Vision* (2009), pp. 145–149
5. C.C. Queirolo, L. Silva, O.R.P. Bellon, M. Pamplona Segundo, 3D face recognition using simulated annealing and the surface interpenetration measure. IEEE Trans. Pattern Anal. Mach. Intell. **32**(1–2), 206–219 (2010)
6. S. Ganguly, D. Bhattacharjee, M. Nasipuri, Fuzzy matching of edge and curvature based features from range images for 3D face recognition. Intell. Autom. Soft Comput. **23**(1), 51–62 (2016)
7. T. Terada, Y. Chen, R. Kimura, 3D facial landmark detection using deep convolutional neural networks, in *2018 14th International Conference on Natural Computation, Fuzzy Systems and Knowledge Discovery (ICNC-FSKD)* (2018), pp. 390–393
8. G. Lee, M. Kwon, S.K. Sri, M. Lee, Emotion recognition based on 3D fuzzy visual and EEG features in movie clips. Neurocomputing **144**, 560–568 (2014)
9. K. Yurtkan, H. Demirel, Feature selection for improved 3D facial expression recognition. Pattern Recogn. Lett. **38**, 26–33 (2014)
10. R. Karthiga, S. Mangai, Feature selection using multi-objective modified genetic algorithm in multimodal biometric system. J. Med. Syst. **43**(7), 214 (2019)
11. G. Amirthalingam, G. Radhamani, New chaff point based fuzzy vault for multimodal biometric cryptosystem using particle swarm optimization. J. King Saud Univ. Comput. Inf. Sci. **28**(4), 381–394 (2016)
12. A. Kumar, M. Hanmandlu, H. Gupta, Ant colony optimization based fuzzy binary decision tree for bimodal hand knuckle verification system. Expert Syst. Appl. **40**(2), 439–449 (2013)
13. L. Dora, S. Agrawal, R. Panda, A. Abraham, An evolutionary single Gabor kernel based filter approach to face recognition. Eng. Appl. Artif. Intell. **62**, 286–301 (2017)

14. O. Zanganeh, B. Srinivasan, N. Bhattacharjee, Partial fingerprint matching through region-based similarity, in *2014 International Conference on Digital Image Computing: Techniques and Applications (DICTA)* (2014), pp. 1–8
15. N. Ahmed, A. Varol, Minutiae based partial fingerprint registration and matching method, in *2018 6th International Symposium on Digital Forensic and Security (ISDFS)* (2018), pp. 1–5
16. S. Huang, Z. Zhang, Y. Zhao, J. Dai, C. Chen, Y. Xu, E. Zhang, L. Xie, 3D fingerprint imaging system based on full-field fringe projection profilometry. Opt. Lasers Eng. **52**, 123–130 (2014)
17. F. Liu, D. Zhang, L. Shen, Study on novel curvature features for 3D fingerprint recognition. Neurocomputing **168**, 599–608 (2015)
18. F. Liu, D. Zhang, 3D fingerprint reconstruction system using feature correspondences and prior estimated finger model. Pattern Recogn. **47**(1), 178–193 (2014)
19. C. Lin, A. Kumar, Contactless and partial 3D fingerprint recognition using multi-view deep representation. Pattern Recogn. **83**, 314–327 (2018)
20. G.K.O. Michael, T. Connie, A.B.J. Teoh, A contactless biometric system using multiple hand features. J. Vis. Commun. Image Represent. **23**(7), 1068–1084 (2012)
21. J.J. Winston, D.J. Hemanth, A comprehensive review on iris image-based biometric system. Soft Comput. **23**(19), 9361–9384 (2019)
22. Y. Ran Zhai, J. Zhong, Y. Ran, K. Li, D. Zeng, A novel method of obtaining 3D images of detached retina. Comput. Methods Prog. Biomed. **108**(2), 665–668 (2012)
23. F. Cohen, S. Sowmithran, C. Li, Iris identification in 3D, in *Image Analysis* (Springer International Publishing, Cham, 2019), pp. 324–335
24. M.S. Khan, R. Malik, A. Siddique, A. Nawaz, A new 3D eyeball tracking system to enhance the usability of page scrolling. Optik **185**, 1270–1276 (2019)
25. A. Alsubari, P. Lonkhande, R.J. Ramteke, Fuzzy-based classification for fusion of palmprint and iris biometric traits, in *Recent Trends in Signal and Image Processing*. Advances in Intelligent Systems and Computing, vol. 922, 2019
26. P. Ramamoorthy, R. Gayathri, Feature level fusion of palmprint and iris. Int. J. Comput. Sci. Issues **9**(1), 194–203 (2012)
27. R. Álvarez Mariño, F.H. Álvarez, L.H. Encinas, A crypto-biometric scheme based on iris-templates with fuzzy extractors. Inf. Sci. **195**, 91–102 (2012)
28. X. Zhou, C. Busch, Measuring privacy and security of iris fuzzy commitment, in *2012 IEEE International Carnahan Conference on Security Technology (ICCST)* (2012), pp. 168–173
29. R. Subban, N. Susitha, D.P. Mankame, Efficient iris recognition using Haralick features based extraction and fuzzy particle swarm optimization. Clust. Comput. **21**(1), 79–90 (2018)
30. K. Roy, P. Bhattacharya, C.Y. Suen, Towards nonideal iris recognition based on level set method, genetic algorithms and adaptive asymmetrical SVMS. Eng. Appl. Artif. Intell. **24**(3), 458–475 (2011)
31. S. Marcel, M.S. Nixon, S.Z. Li, *Handbook of Biometric Anti-Spoofing*, vol. 1 (Springer, London, 2014)
32. J. Galbally, S. Marcel, J. Fierrez, Biometric antispoofing methods: a survey in face recognition. IEEE Access **2**, 1530–1552 (2014)
33. A. Jourabloo, Y. Liu, X. Liu, Face de-spoofing: anti-spoofing via noise modeling, Lecture Notes in Computer Science, 2018, pp. 297–315
34. I. Chingovska, A. Anjos, S. Marcel, On the effectiveness of local binary patterns in face anti-spoofing, in *2012 BIOSIG – Proceedings of the International Conference of Biometrics Special Interest Group (BIOSIG)* (2012), pp. 1–7
35. Y. Tang, X. Wang, X. Jia, L. Shen, Fusing multiple deep features for face anti-spoofing, in *Biometric Recognition*, ed. by J. Zhou, Y. Wang, Z. Sun, Z. Jia, J. Feng, S. Shan, K. Ubul, Z. Guo (Springer, 2018), pp. 321–330
36. J. Yang, D. Schonfeld, Virtual focus and depth estimation from defocused video sequences. IEEE Trans. Image Process. **19**(3), 668–679 (2010)
37. Y. Wang, F. Nian, T. Li, Z. Meng, K. Wang, Robust face anti-spoofing with depth information. J. Vis. Commun. Image Represent. **49**, 332–337 (2017)

38. X. Tu, Y. Fang, Ultra-deep neural network for face anti-spoofing, in *Neural Information Processing. ICONIP 2017*, ed. by D. Liu, S. Xie, Y. Li, D. Zhao, E.S. El-Alfy. Lecture Notes in Computer Science, vol. 10635 (Springer, Cham, 2017), pp. 686–695

39. L.-B. Zhang, F. Peng, L. Qin, M. Long, Face spoofing detection based on color texture Markov feature and support vector machine recursive feature elimination. J. Vis. Commun. Image Represent. **51**, 56–69 (2018)

40. B. Hamdan, K. Mokhtar, A self-immune to 3D masks attacks face recognition system. Signal Image Video Process. **12**(6), 1053–1060 (2018)

41. N. Erdogmus, S. Marcel, Spoofing in 2D face recognition with 3D masks, in *2013 International Conference of the BIOSIG Special Interest Group (BIOSIG)*, Darmstadt, 2013, pp. 1–8

42. M.P. Beham, S.M.M. Roomi, Anti-spoofing enabled face recognition based on aggregated local weighted gradient orientation. Signal Image Video Process. **12**(3), 531–538 (2018)

43. B. Hamdan, K. Mokhtar, The detection of spoofing by 3D mask in a 2D identity recognition system. Egyptian Inf. J. **19**(2), 75–82 (2018)

44. P. Kavitha, K. Vijaya, Optimal feature-level fusion and layered k-support vector machine for spoofing face detection. Multimed. Tools Appl. **77**(20), 26509–26543 (2018)

45. J. Guo, X. Zhu, J. Xiao, Z. Lei, G. Wan, S.Z. Li, Improving face anti-spoofing by 3D virtual synthesis, 2019, arXiv preprint arXiv:1901.00488

46. Z. Xia, C. Yuan, R. Lv, X. Sun, N.N. Xiong, Y. Shi, A novel weber local binary descriptor for fingerprint liveness detection. IEEE Trans. Syst. Man Cybern. Syst. 1–11 (2018)

47. R.K. Dubey, J. Goh, V.L.L. Thing, Fingerprint liveness detection from single image using low-level features and shape analysis. IEEE Trans. Inf. Forensics Secur. **11**(7), 1461–1475 (2016)

48. R.F. Nogueira, R. de Alencar Lotufo, R. Campos Machado, Fingerprint liveness detection using convolutional neural networks. IEEE Trans. Inf. Forensics Secur. **11**(6), 1206–1213 (2016)

49. A. Krizhevsky, I. Sutskever, G.E. Hinton, Imagenet classification with deep convolutional neural networks, in *Proceedings of the 25th International Conference on Neural Information Processing Systems – Volume 1, NIPS'12* (2012), pp. 1097–1105

50. O. Russakovsky, J. Deng, H. Su, J. Krause, S. Satheesh, S. Ma, Z. Huang, A. Karpathy, A. Khosla, M. Bernstein, A.C. Berg, L. Fei-Fei, Imagenet large scale visual recognition challenge. Int. J. Comput. Vis. **115**(3), 211–252 (2015)

51. J. Galbally, J. Ortiz-Lopez, J. Fierrez, J. Ortega-Garcia, Iris liveness detection based on quality related features, in *2012 5th IAPR International Conference on Biometrics (ICB)* (2012), pp. 271–276

52. K.B. Raja, R. Raghavendra, C. Busch, Presentation attack detection using laplacian decomposed frequency response for visible spectrum and near-infra-red iris systems, in *2015 IEEE 7th International Conference on Biometrics Theory, Applications and Systems (BTAS)* (2015), pp. 1–8

53. A. Czajka, Pupil dynamics for iris liveness detection. IEEE Trans. Inf. Forensics Secur. **10**(4), 726–735 (2015)

54. J.C. Klontz, B.F. Klare, S. Klum, A.K. Jain, M.J. Burge, Open source biometric recognition, in *2013 IEEE Sixth International Conference on Biometrics: Theory, Applications and Systems (BTAS)* (2013), pp. 1–8

55. The Open Source Biometrics Project, Openebts, Openbiometricsinitiative, in http://www.openbiometricsinitiative.org/index.html (2019), pp. 1–7

56. Biometrices at TELECOM SudParis, Biosecure biometrics for secure authentication, in http://biometrics.it-sudparis.eu (2007)

57. A. Mayoue, D. Petrovska-Delacrétaz, Open source reference systems for biometric verification of identity, in *Open Source Development, Communities and Quality* (2008), pp. 397–404

58. N. Fingerprint, Fingerprint, in https://www.nist.gov/programs-projects/fingerprint (2019)

59. Center for Biometrics and Security Research, CASIA iris image database, in http://www.cbsr.ia.ac.cn/IrisDatabase.htm (2005)

60. E. González Agulla, E. Otero Muras, J.L. Alba Castro, C. García Mateo, An open source java framework for biometric web authentication based on bioapi, in *Knowledge-Based Intelligent Information and Engineering Systems* (2007), pp. 809–815

Index

A

Access control
  accountability, 274
  account authentication, 274
  authorization, 274
  definition, 267
  identification, 274
  types, 275
  user's roles, 268
Account authentication control, 274, 277
Additive white Gaussian noise (AWGN)
      channel
  ASK, 101, 102
  BPSK, 104, 105
  FSK, 101, 103
  QAM, 101, 104, 107
  QPSK, 101, 104, 106
Alpha wave, 235
Alzheimer's disorder (AD), 291
Amplitude shift keying (ASK), 101, 102
Angular radial transformation (ART) method,
      341
Atherosclerotic plaques, 165
Attribute-based access control (ABAC), 275
Auto-encoder method, 302

B

Behavioral traits
  keystrokes, 311
  signature, 311
  voice, 310–311
Behavior-based detection approach, 30
Bell-LaPadula model, 270
Benign software

accuracy, 61–62
confusion matrix, 61
k-nearest neighbors, 61
precision, 62
random forest, 61
receiver operating characteristic, 62
recall, 62
Beta wave, 235
Biba security model
  access modes, 276
  advantages and disadvantages, 276–277
  discretionary policies, 276
  mandatory policies, 276
Binary code, 136
Binary phase shift key (BPSK), 104, 105
Bins, 198
Biometric templates, 317
Bit error rate (BER)
  digital modulation techniques, 114–115
  GNU radio, 93
Black hat hacker, 38–39

C

Canny filtering process, 75
Central sensory system (CNS), 164
Classifiers
  classification problems, 119
  cognitive principles, 118–119
  McELM, 126
  McFIS, 123–124
  McNN
    architecture, 122
    knowledge measures, 123
    learning strategies, 122–123

© Springer Nature Switzerland AG 2019
G.R. Sinha (ed.), *Advances in Biometrics*,
https://doi.org/10.1007/978-3-030-30436-2

Printed in the United States
By Bookmasters